Mixed Matrix Membranes

Mixed Matrix Membranes

Special Issue Editor
Clara Casado-Coterillo

MDPI • Basel • Beijing • Wuhan • Barcelona • Belgrade

MDPI

Special Issue Editor
Clara Casado-Coterillo
Universidad de Cantabria
Spain

Editorial Office
MDPI
St. Alban-Anlage 66
4052 Basel, Switzerland

This is a reprint of articles from the Special Issue published online in the open access journal *Membranes* (ISSN 2077-0375) from 2018 to 2019 (available at: https://www.mdpi.com/journal/membranes/special_issues/mixed_matrix_mem).

For citation purposes, cite each article independently as indicated on the article page online and as indicated below:

LastName, A.A.; LastName, B.B.; LastName, C.C. Article Title. *Journal Name* **Year**, *Article Number*, Page Range.

ISBN 978-3-03921-976-6 (Pbk)
ISBN 978-3-03921-977-3 (PDF)

Cover image courtesy of Clara Casado-Coterillo.

Contents

About the Special Issue Editor vii

Preface to "Mixed Matrix Membranes" ix

Clara Casado-Coterillo
Mixed Matrix Membranes
Reprinted from: *Membranes* 2019, 9, 149, doi:10.3390/membranes9110149 1

Mahdi Ahmadi, Saravanan Janakiram, Zhongde Dai, Luca Ansaloni and Liyuan Deng
Performance of Mixed Matrix Membranes Containing Porous Two-Dimensional (2D) and
Three-Dimensional (3D) Fillers for CO_2 Separation: A Review
Reprinted from: *Membranes* 2018, 8, 50, doi:10.3390/membranes8030050 6

Clara Casado-Coterillo, Ana Fernández-Barquín, Susana Valencia and Ángel Irabien
Estimating CO_2/N_2 Permselectivity through Si/Al = 5 Small-Pore Zeolites/PTMSP Mixed
Matrix Membranes: Influence of Temperature and Topology
Reprinted from: *Membranes* 2018, 8, 32, doi:10.3390/membranes8020032 54

Sandra Sánchez-González, Nazely Diban and Ane Urtiaga
Hydrolytic Degradation and Mechanical Stability of Poly(ε-Caprolactone)/Reduced Graphene
Oxide Membranes as Scaffolds for In Vitro Neural Tissue Regeneration
Reprinted from: *Membranes* 2018, 8, 12, doi:10.3390/membranes8010012 69

**Gabriel Guerrero, May-Britt Hägg, Christian Simon, Thijs Peters, Nicolas Rival and
Christelle Denonville**
CO_2 Separation in Nanocomposite Membranes by the Addition of Amidine and Lactamide
Functionalized POSS® Nanoparticles into a PVA Layer
Reprinted from: *Membranes* 2018, 8, 28, doi:10.3390/membranes8020028 83

Muntazim Munir Khan, Sergey Shishatskiy and Volkan Filiz
Mixed Matrix Membranes of Boron Icosahedron and Polymers of Intrinsic Microporosity
(PIM-1) for Gas Separation
Reprinted from: *Membranes* 2018, 8, 1, doi:10.3390/membranes8010001 100

**Parashuram Kallem, Christophe Charmette, Martin Drobek, Anne Julbe, Reyes Mallada and
Maria Pilar Pina**
Exploring the Gas-Permeation Properties of Proton-Conducting Membranes Based on Protic
Imidazolium Ionic Liquids: Application in Natural Gas Processing
Reprinted from: *Membranes* 2018, 8, 75, doi:10.3390/membranes8030075 118

About the Special Issue Editor

Clara Casado-Coterillo completed her Ph.D. in chemical engineering at the age of 27 years from the University of Cantabria, Spain, where she came back as senior researcher in 2012, after pursuing knowledge on the synthesis and characterization of different membrane materials for diverse molecular separation applications in international research centers as the University of Hiroshima (Japan), the University of Twente (The Netherlands), the Institut Européen des Membranes (France), the University of Zaragoza and the Institute of Chemical Technology (Spain). She has written 52 publications in JCR-indexed journals that have been cited over 1101 times, her publication H-index is 20 and has participated in 20 projects in competitive calls, 7 as principal investigator, and directed 2 Ph.D. international thesis (1 on going).

Preface to "Mixed Matrix Membranes"

This Special Issue was motivated by the gap between a growing interest in developing novel mixed matrix membranes by various research groups and the lack of large-scale implementation. It contains six publications dealing with actual opportunities that materials science development offers to overcome the challenges of mixed matrix membrane fabrication for their application as solutions in environmental and health issues of the society of 21st century.

Clara Casado-Coterillo
Special Issue Editor

membranes MDPI

Editorial

Mixed Matrix Membranes

Clara Casado-Coterillo

Department of Chemical and Biomolecular Engineering, E.T.S. Ingenieros Industriales y Telecomunicación, Universidad de Cantabria, Av. Los Castros, s/n, 39005 Santander, Cantabria, Spain; casadoc@unican.es; Tel.: +34-942-206777

Received: 5 November 2019; Accepted: 7 November 2019; Published: 10 November 2019

Abstract: In recent decades, mixed matrix membranes (MMMs) have attracted considerable interest in research laboratories worldwide, motivated by the gap between the growing interest in developing novel mixed matrix membranes by various research groups and the lack of large-scale implementation. This Special Issue contains six publications dealing with the current opportunities and challenges of mixed matrix membranes development and applications as solutions for the environmental and health challenges of 21st century society.

Keywords: membrane fabrication; membrane modification; flat-sheet membrane, characterization techniques; hollow fiber membrane; filler dispersion; compatibility; gas separation; ion exchange capacity; water vapor

1. Introduction

This Special Issue, entitled "Mixed Matrix Membranes", was motivated by the observed gap between the growing interest of research laboratories in developing novel mixed matrix membranes (MMMs) and the lack of large-scale implementation. MMMs, consisting of the mixing of innovative fillers and processable polymer matrices, may fill in this gap for conventional membranes to address industrial process intensifications challenges [1]. The papers compiled within this Special issue can be read as single chapters of a global story orientated toward the advancement of mixed matrix membranes and novel materials in membrane technology in response to some technical challenges faced by chemical industries and society, from CO_2 capture and utilization to tissue engineering applications in biomedicine. They are all connected through important issues regarding fabrication, such as compatibility and adhesion, the effect of porous and non-porous fillers on the polymer matrices, types of additives/fillers (zeolites, ionic liquids, ion-exchange materials, layered porous materials, metal organic frameworks (MOFs), etc.), and characterization (e.g., chemical, structural, morphological, electrical, compositional, mechanical and topographical properties, as well as membrane transport and separation).

2. Highlights of the Special Issue

The papers included in this special issue direct the developments in MMMs to some of the major challenges faced by society in the 21st century, mainly CO_2 separation from other gases as a way in which to tackle climate change, and biomedical applications. One of the most important aspects is thus the selection of the appropriate material for both the matrix and dispersed phases to eliminate non-ideal morphologies created at their interfaces [1]. With these aims, several kinds of membranes have been addressed, as will be presented in the following paragraphs.

2.1. Mixed Matrix Membranes with Porous Fillers

The Special Issue opens with a review presenting a complete synopsis of the inherent capacities of several porous nanofillers, distinguishing between two-dimensional (2D) and three-dimensional

(3D) shaped fillers [2] for CO_2 separation from other gases. Gas permeation performances of selected hybrids with 3D fillers and porous nanosheets have been summarized and discussed with respect to each type and the effects of their embedment in polymers to make mixed matrix membranes for the separation of CO_2 from other gases [3]. The particular challenge of achieving an intimate adhesion between fillers and polymer matrices to avoid the presence of defects and assure a correct synergy of the new MMM material is addressed by the studies of metal organic frameworks (MOFs) [4], and porous organic frameworks (POFs) [5,6], in consideration of their organic nature and high CO_2 uptake properties. The oldest studied MMMs with porous fillers and glassy polymers for gas separation are zeolite–polymer membranes. The additional porosity provide additional transport mechanisms that account for their non-ideal performance [7]. The prediction of the mixed matrix membrane permeability and selectivity has been explored by different morphological models that have been thoroughly reviewed [8]. The feature paper contained in this Special Issue compares several of those models regarding the effect of filler type and topology on CO_2 and N_2 permeability using zeolites of different topologies (CHA, RHO, and LTA) and identical Si/Al compositional ratio, embedded in a high permeability glassy polymer, poly(trimethylsilyl-1-propyne) (PTMSP), as a function of temperature, zeolite loading, and topology [9]. The evolution of temperature and its influence on non-idealities, such as membrane rigidification, zeolite–polymer compatibility, and sieve pore blockage, allow prediction of the structure-performance relationship for further membrane development for the first time [10].

The recent advances in the synthesis and improvements of 2D and 3D porous nanophases have driven continuous research within the development of MMMs for gas separation purposes. In particular, the possibility of tuning the pore diameter to a gas-sieving level and the CO_2-philicity of the pore cavity has the potential to facilitate the simultaneous enhancement of the solubility and diffusivity coefficient of carbon dioxide and reduced CO_2 plasticization when high pressures are necessary [11,12]. Therefore, CO_2 permeability and selectivity can be expected to benefit from these features, leading to a shift in the separation performance towards the upper right corner of the Robeson plot as a function also of the rubbery or glassy nature of the polymer matrix [13].

2D porous fillers offer a high surface area to volume ratio that provides higher contact between the filler and the polymer matrix in comparison with other filler morphologies [14]. This may result in the development of new applications, such as those explored by Sanchez-Gonzalez et al. [15] in this Special Issue. Their paper aims at elucidating the applicability of poly(caprolactone) (PCL) and reduced graphene oxide (rGO) MMMs as scaffolds for in vitro neural regeneration, by correlating the morphological, chemical, and differential scanning calorimetry (DSC) results with the membrane performance under simulated in vitro culture conditions (phosphate buffer solution (PBS) at 37 °C) for 1 year. The high internal porosity of the membranes facilitated water permeation and resulted in an accelerated hydrolytic degradation and molecular weight reduction. The presence of the rGO nanoplatelets caused the pH to be barely affected, while accelerating the loss of mechanical stability of the membranes. However, it is envisioned that the gradual degradation of the PCL/rGO membranes could facilitate cells infiltration, interconnectivity, and tissue formation. The relationship between structure and function seems again highly important in the opening up of novel applications for MMMs.

2.2. Mixed Matrix Membranes Filled with Nonporous Fillers

Membranes must offer a high CO_2 permeability in order to compete with conventional membranes or other separation processes in CO_2 capture and climate change mitigation processes [16]. Organic–inorganic nanocomposite membranes resulting from the in situ generation of inorganic nanoparticles in the polymer matrix can offer much higher gas permeabilities with similar selectivities than MMMs prepared by dispersion of inorganic fillers in the polymer matrix [17]. In this Special Issue, Guerrero et al. [18] present two differently functionalized types of polyhedral oligomeric silsesquioxanes (POSS®) nanoparticles as additives for nanocomposite membranes for CO_2 separation. Composite membranes were produced by casting a polyvinyl alcohol (PVA) layer, containing the functionalized POSS® nanoparticles, on a polysulfone (PSf) porous support. The compatibility between the nanoparticles and the polymer was observed by FTIR. Differential scanning calorimetry (DSC) and dynamic mechanical analysis (DMA) show an increment of the crystalline regions affected by the conformation of the polymer chains, decreasing the gas separation performance. Moreover, these nanocomposite membranes did not show separation according to a facilitated transport mechanism, as might be expected based on their functionalized amino-groups; thus, solution-diffusion was the main mechanism responsible for the transport phenomena [19].

Tuning the polymer free volume available for transport by disrupting the polymer chain packing with nanosized particles also has an effect on the gas permeation and stability of highly permeable rigid polymers [20]. Khan et al. [21] proposed here yet another nanofiller, potassium dodecahydrododecaborate ($K_2B_{12}H_{12}$)—a polynuclear borane with potential in materials science and biomedicine—as a new filler to be added to the rigid structure of PIM-1 in order to improve its gas permeation properties and robustness [22]. Although the permeability performance of the prepared MMMs mainly depended on the addition of nanofillers rather than the effect of interfacial zone and the O_2/N_2 separation factor was almost constant for all the membranes, overall increases in permeability and diffusivity were observed for all tested gases coupled with the reduction in gas pair selectivity.

2.3. Mixed Matrix Membranes Filled with Ionic Liquids

The search for a good adhesion between polymers and fillers has also been directed to ionic liquids (ILs). ILs have been thoroughly explored in the last few decades as an alternative form of solvent to amines in CO_2 separation processes in supported ionic liquid membranes because of several advantages, such as negligible vapor pressure [16]. The combination of ionic liquids into a polymer matrix is an approach to enhance the mechanical stability of the separation process by avoiding working with a fluid phase [23,24]. This Special Issue presents an experimental study exploring the potential of supported ionic liquid membranes (SILMs) prepared by infiltration of protic imidazolium ionic liquids (ILs) into randomly nanoporous polybenzimidazole (PBI) membranes for CH_4/N_2 separation [25]. The polymerization, monitored by Fourier transform infrared (FTIR) spectroscopy, and the concentration of the protic entities in the membranes evaluated by thermogravimetric analysis (TGA) were correlated to the gas permeability values of N_2 and CH_4 at 313 K, 333 K, and 363 K in terms of the preferential cavity formation and favorable solvation of methane in the apolar domains of the protic ionic network. The transport mechanism of the as-prepared SILMs is solubility-dominated at high temperature, which can be compared with MMMs of similar components [26].

3. Final Remarks

Overall, the editor is convinced that mixed matrix membranes have a lot more to contribute than what has already been demonstrated worldwide. It is hoped that readers enjoy this Special Issue and gain inspiration from it for their own work. In the end, technological changes are the fruit of ideas planted as seeds in researchers' minds: the more that individual minds are connected to each other, the higher the probability of creating originality. In this sense, this Special Issue represents a small attempt to increase the connectivity among interested minds, regarding the contributions to solve technological challenges in mixed matrix membrane development, and it shows the possibilities of synergies that the combination of compatible fillers and polymers can offer to environmental and health issues faced by society in the 21st century.

Funding: Financial support by the Spanish Ministry for Science and Universities under project grant no. CTQ2016-76231-C2-1-R at the Universidad de Cantabria is gratefully acknowledged.

Acknowledgments: The editor acknowledges all the contributors to this Special Issue and thanks them for generously taking the time and effort to prepare a manuscript.

Conflicts of Interest: The editor declares no conflict of interest. The funders had no role in the design of the study; in the collection, analyses, or interpretation of data; in the writing of the manuscript, or in the decision to publish the results.

References

1. Ebadi Amooghin, A.; Mashhadikhan, S.; Sanaeepur, H.; Moghadassi, A.; Matsuura, T.; Ramakrishna, S. Substantial breakthroughs on function-led design of advanced materials used in mixed matrix membranes (MMMs): A new horizon for efficient CO_2 separation. *Prog. Mater. Sci.* **2019**, *102*, 222–295. [CrossRef]
2. Ahmadi, M.; Janakiram, S.; Dai, Z.; Ansaloni, L.; Deng, L. Performance of mixed matrix membranes containing porous two-dimensional (2D) and three-dimensional (3D) fillers for CO_2 separation: A review. *Membranes* **2018**, *8*, 50. [CrossRef] [PubMed]
3. Gascon, J.; Kapteijn, F.; Zornoza, B.; Sebastián, V.; Casado, C.; Coronas, J. Practical approach to zeolitic membranes and coatings: State of the art, opportunities, barriers, and future perspectives. *Chem. Mater.* **2012**, *24*, 2829–2844. [CrossRef]
4. Sabetghadam, A.; Liu, X.; Gottmer, S.; Chu, L.; Gascon, J.; Kapteijn, F. Thin mixed matrix and dual layer membranes containing metal-organic framework nanosheets and Polyactive^{TM} for CO_2 capture. *J. Membr. Sci.* **2019**, *570–571*, 226–235. [CrossRef]
5. Kang, Z.; Peng, Y.; Qian, Y.; Yuan, D.; Addicoat, M.A.; Heine, T.; Hu, Z.; Tee, L.; Guo, Z.; Zhao, D. Mixed Matrix Membranes (MMMs) Comprising Exfoliated 2D Covalent Organic Frameworks (COFs) for Efficient CO_2 Separation. *Chem. Mater.* **2016**, *28*, 1277–1285. [CrossRef]
6. Shan, M.; Seoane, B.; Andres-garcia, E.; Kapteijn, F.; Gascon, J. Mixed-matrix membranes containing an azine-linked covalent organic framework: In fl uence of the polymeric matrix on post-combustion CO_2-capture. *J. Membr. Sci.* **2018**, *549*, 377–384. [CrossRef]
7. Mahajan, R.; Burns, R.; Schaeffer, M.; Koros, W.J. Challenges in forming successful mixed matrix membranes with rigid polymeric materials. *J. Appl. Polym. Sci.* **2002**, *86*, 881–890. [CrossRef]
8. Vinh-thang, H.; Kaliaguine, S. Predictive Models for Mixed-Matrix Membrane Performance: A Review. *Chem. Rev.* **2013**, *113*, 4080–5028. [CrossRef]
9. Fernández-Barquín, A.; Casado-Coterillo, C.; Palomino, M.; Valencia, S.; Irabien, A. Permselectivity improvement in membranes for CO_2/N_2 separation. *Sep. Purif. Technol.* **2016**, *157*, 102–111. [CrossRef]
10. Shen, Y.; Lua, A.C. Theoretical and Experimental Studies on the Gas Transport Properties of Mixed Matrix Membranes Based on Polyvinylidene Fluoride. *AIChE J.* **2014**, *59*, 4715–4726. [CrossRef]
11. Gong, H.; Lee, S.S.; Bae, T.H. Mixed-matrix membranes containing inorganically surface-modified 5A zeolite for enhanced CO_2/CH_4 separation. *Microporous Mesoporous Mater.* **2017**, *237*, 82–89. [CrossRef]
12. Shahid, S.; Nijmeijer, K. Performance and plasticization behavior of polymer—MOF membranes for gas separation at elevated pressures. *J. Membr. Sci.* **2014**, *470*, 166–177. [CrossRef]
13. Robeson, L.M.; Liu, Q.; Freeman, B.D.; Paul, D.R. Comparison of transport properties of rubbery and glassy polymers and the relevance to the upper bound relationship. *J. Membr. Sci.* **2015**, *476*, 421–431. [CrossRef]
14. Zornoza, B.; Gorgojo, P.; Casado, C.; Téllez, C.; Coronas, J. Mixed matrix membranes for gas separation with special nanoporous fillers. *Desalin. Water Treat.* **2011**, *27*, 42–47. [CrossRef]
15. Sánchez-González, S.; Diban, N.; Urtiaga, A. Hydrolytic degradation and mechanical stability of poly(ε-Caprolactone)/reduced graphene oxide membranes as scaffolds for in vitro neural tissue regeneration. *Membranes* **2018**, *8*, 12. [CrossRef]
16. Wang, M.; Joel, A.S.; Ramshaw, C.; Eimer, D.; Musa, N.M. Process intensification for post-combustion CO_2 capture with chemical absorption: A critical review. *Appl. Energy* **2015**, *158*, 275–291. [CrossRef]
17. Madhavan, K.; Reddy, B.S.R. Structure-gas transport property relationships of poly(dimethylsiloxane-urethane) nanocomposite membranes. *J. Membr. Sci.* **2009**, *342*, 291–299. [CrossRef]
18. Guerrero, G.; Hägg, M.B.; Simon, C.; Peters, T.; Rival, N.; Denonville, C. CO_2 separation in nanocomposite membranes by the addition of amidine and lactamide functionalized POSS® nanoparticles into a PVA layer. *Membranes* **2018**, *8*, 28. [CrossRef]
19. Wijmans, J.G.; Baker, R.W. The solution-diffusion model: A review. *J. Membr. Sci.* **1995**, *107*, 1–21. [CrossRef]
20. Hill, A.J.; Freeman, B.D.; Jaffe, M.; Merkel, T.C.; Pinnau, I. Tailoring nanospace. *J. Mol. Struct.* **2005**, *739*, 173–178. [CrossRef]
21. Khan, M.M.; Shishatskiy, S.; Filiz, V. Mixed matrix membranes of boron icosahedron and polymers of intrinsic microporosity (PIM-1) for gas separation. *Membranes* **2018**, *8*, 1. [CrossRef] [PubMed]
22. Chen, X.Y.; Vinh-Thang, H.; Ramirez, A.A.; Rodrigue, D.; Kaliaguine, S. Membrane gas separation technologies for biogas upgrading. *RSC Adv.* **2015**, *5*, 24399–24448. [CrossRef]

23. Dai, Z.; Noble, R.D.; Gin, D.L.; Zhang, X.; Deng, L. Combination of ionic liquids with membrane technology: A new approach for CO_2 separation. *J. Membr. Sci.* **2016**, *497*, 1–20. [CrossRef]

24. Santos, E.; Rodríguez-Fernández, E.; Casado-Coterillo, C.; Irabien, A. Hybrid ionic liquid-chitosan membranes for CO_2 separation: Mechanical and thermal behavior. *Int. J. Chem. React. Eng.* **2016**, *14*, 713–718. [CrossRef]

25. Kallem, P.; Charmette, C.; Drobek, M.; Julbe, A.; Mallada, R.; Pina, M.P. Exploring the gas-permeation properties of proton-conducting membranes based on protic imidazolium ionic liquids: Application in natural gas processing. *Membranes* **2018**, *8*, 75. [CrossRef]

26. Rowe, B.W.; Robeson, L.M.; Freeman, B.D.; Paul, D.R. Influence of temperature on the upper bound: Theoretical considerations and comparison with experimental results. *J. Membr. Sci.* **2010**, *360*, 58–69. [CrossRef]

membranes

MDPI

Review

Performance of Mixed Matrix Membranes Containing Porous Two-Dimensional (2D) and Three-Dimensional (3D) Fillers for CO_2 Separation: A Review

Mahdi Ahmadi, Saravanan Janakiram, Zhongde Dai, Luca Ansaloni * and Liyuan Deng *

Department of Chemical Engineering, Norwegian University of Science and Technology (NTNU), NO-7491 Trondheim, Norway; mahdi.ahmadi@ntnu.no (M.A.); saravanan.janakiram@ntnu.no (S.J.); zhongde.dai@ntnu.no (Z.D.)
* Correspondence: luca.ansaloni@ntnu.no (L.A.); liyuan.deng@ntnu.no (L.D.); Tel.: +47-7359-4112 (L.D.)

Received: 26 June 2018; Accepted: 22 July 2018; Published: 28 July 2018

Abstract: Application of conventional polymeric membranes in CO_2 separation processes are limited by the existing trade-off between permeability and selectivity represented by the renowned upper bound. Addition of porous nanofillers in polymeric membranes is a promising approach to transcend the upper bound, owing to their superior separation capabilities. Porous nanofillers entice increased attention over nonporous counterparts due to their inherent CO_2 uptake capacities and secondary transport pathways when added to polymer matrices. Infinite possibilities of tuning the porous architecture of these nanofillers also facilitate simultaneous enhancement of permeability, selectivity and stability features of the membrane conveniently heading in the direction towards industrial realization. This review focuses on presenting a complete synopsis of inherent capacities of several porous nanofillers, like metal organic frameworks (MOFs), Zeolites, and porous organic frameworks (POFs) and the effects on their addition to polymeric membranes. Gas permeation performances of select hybrids with these three-dimensional (3D) fillers and porous nanosheets have been summarized and discussed with respect to each type. Consequently, the benefits and shortcomings of each class of materials have been outlined and future research directions concerning the hybrids with 3D fillers have been suggested.

Keywords: mixed matrix membranes; CO_2 separation; porous nanoparticles

1. Introduction

An wide scientific consensus is nowadays established in the international community over the anthropogenic climate change and global warming due to a drastic increase of atmospheric level of CO_2 [1]. Anthropogenic activities within transportation, energy supply from fossil fuels [2], and raw materials (e.g., cement, steel) production [3] have significantly contributed to increase in levels of CO_2 emissions over the last century, raising the CO_2 concentration in the atmosphere [4]. The primary strategy to mitigate CO_2 emission in the short term is carbon capture and sequestration (CCS), which mainly includes post-combustion (capture downstream to the combustion), oxy-fuel (purified O_2 used for the combustion), and pre-combustion (capture upstream to the combustion) processes [2]. Furthermore, CO_2 separation is relevant also for other applications, such as Natural Gas sweetening, where acid components in the presence of water can corrode pipelines and equipment, thus lowering the value of the natural gas [3,5]. Therefore, the development of efficient technologies to separate and capture CO_2 is of primary interest.

Physical and chemical adsorption/absorption technologies have been widely applied to industrial plants to separate CO_2 from gaseous streams. These conventional methods exploit pressure and temperature swing absorption/adsorption, which are typically energy-intensive and are not preferred from an environmental and economic standpoint [6]. The most mature technology for post combustion application is absorption using amine-base solvents, but, despite the efforts that are made, the increase in the cost of electricity would be still above the limit of 35%, which is identified as viable solution from a market perspective [7]. When compared to traditional technologies, membrane-based gas separation technology offers several advantages: lower energy consumption (no need for regeneration), no use of harmful chemicals, modularity and easier scalability. Additionally, membrane gas separation offers lower capital and operating costs. Depending on their base material, membranes used for CO_2 separation can be separated in inorganic or polymeric. Even though inorganic membranes offer good separation abilities, polymeric materials are preferred for the application that requires large separation area, due to the lower production costs and easier processability. However, constant research is ongoing in order to improve the state-of-the-art separation for polymeric membranes, aiming at improving their competitiveness to traditional technologies.

Gas transport through a nonporous polymeric membrane is typically based on the "solution-diffusion" mechanism. Conceptually, the gas molecules is absorbed on the upstream side of the membrane layer, it diffuses across the thickness, and is finally desorbed on the downstream side. The permeation is therefore described as contribution of a thermodynamic parameter (solubility) and a kinetic factor (diffusivity), which affect the transport of gas molecules across the membrane matrix. The two most important features characterizing gas permeation membranes are permeability and selectivity [8]. Permeability of a given gaseous species (A) is as an intrinsic property of the material and is defined as the specific flux (J_A) normalized on the membrane thickness (ℓ) and partial pressure difference between the upstream and downstream side of the membrane (Δp_A), as showed in Equation (1):

$$P_A = \frac{J_A \cdot \ell}{\Delta p_A} \qquad (1)$$

Permeability is frequently reported in Barrer (1 Barrer = 10^{-10} cm^3 (STP) cm^{-1} s^{-1} cmHg^{-1} = 3.346×10^{-16} mol m^{-1} Pa^{-1} s^{-1}). For the implementation of membranes in real process operations, membranenologists have to focus on the fabrication of thin composite membranes, aiming at maximizing the transmembrane flux of permeants [9]. In this perspective, the capacity of a membrane to allow for a specific gas to permeate through the selective layer is described by means of permeance, often reported in GPU (gas permeation unit, 1 GPU = 10^{-6} cm^3 (STP) cm^{-2} s^{-1} cmHg^{-1} = 3.346×10^{-10} mol m^{-2} Pa^{-1} s^{-1}). Unlike permeability, permeance is not an intrinsic property of the polymeric material, but it directly quantifies the actual transmembrane flux achievable for a given driving force. For this reason, the gas permeance is described as the ratio of the flux (J_A) and the driving force (Δp_A). The other key membrane feature is the separation factor (or selectivity), which is defined as the molar ratio of gases A and B in the permeate (y) and in the feed side (x), with A being the most permeable gaseous species:

$$\alpha = \frac{y_A/y_B}{x_A/x_B} \qquad (2)$$

When single gas tests are performed, the membrane "ideal" selectivity can be estimated as the ratio between the permeability of the two penetrants [10].

The analysis of the performance of a larger amount of polymers for gas permeation allowed for Robeson [11,12] to highlight the existence of a trade-off between permeability and selectivity for materials governed by the solution-diffusion mechanism. This relation between permeability and selectivity reveals that for polymer membranes, an increase in permeability happens typically at the expense of selectivity, and vice versa. In the attempt to provide a more fundamental explanation, of an empirical relationship between permeability and selectivity was established [13,14], and it was shown that in the determination of the upper bound slope, the diffusion coefficient plays a dominant role as compared to the solubility coefficient.

Among the different strategies to overcome the upper bound (fabrication of highly permeable polymers, such as thermally rearranged polymers [15], high free volume glassy polymers [16]; facilitated transport membranes [17]), a promising approach is the embedment of different phases (inorganic or liquid) within the membrane matrix, fabricating so-called hybrid membranes. Inorganic membranes that are made of non-polymeric materials, such as carbon molecular sieves, zeolites, or metal organic frameworks (MOFs) are typically characterized by performance exceeding the upper bound [18], but their cost and poor mechanical stability limit their applicability at large scale. Nevertheless, the dispersion of high performance nano-phases within a polymer matrix can significantly improve the neat polymer separation properties. In recent years, extensive efforts have been made in order to fabricated hybrid materials containing dispersed inorganic phases within polymeric matrices [8,19–21].

Based on the type of the embedded phase, hybrid membranes are classified in two main groups, known as mixed matrix membranes and nanocomposite membranes [10]. Nanocomposite membranes contain nano-sized impermeable nanoparticles that can contribute to the overall transport via surface adsorption or due to the presence of moieties with a specific affinity towards a specific penetrant. In our previous review, a broad overview of the performance of nanocomposite membranes has been presented [22]. On the opposite side, in mixed matrix membranes, the embedded phase contributes to a secondary transport mechanism. The fillers are typically porous and the pore architecture confers a larger CO_2 solubility and/or diffusivity selectivity to the hybrid when compared to the neat polymer. Based on the nature of the embedded phase, the secondary transport mechanism can be described by molecular sieving, surface diffusion, or Knudsen diffusion. Nevertheless, the effect of the fillers on the overall transport through the hybrid membrane is inherently related to the type of polymer-particle interface that is achieved [10]. Ideal adhesion between the two phases would allow for achieving the largest enhancement, whereas poor interface morphology would result in the formation of unselective voids, frequently reflected by deteriorated separation performances.

We previously categorized [22] inorganic fillers in different categories based on their morphology (zero- to three-dimensional morphology), specifying which type constitutes the class of nanocomposite (zero-dimensional (0D) to two-dimensional (2D) nanofillers) or mixed matrix membranes (three-dimensional (3D) nanoparticles). Silica, metal oxide, nanotubes, nanofibers, and graphene derivate are categorized within the nanoparticles used for the fabrication of nanocomposite membranes, whereas zeolites, metal organic frameworks (MOFs), and porous organic frameworks (POFs) are listed as nano-phases that are used for the fabrication of mixed matrix membranes.

The current report mainly focuses on the latest advances in hybrid membranes containing phases that are able to add secondary transport mechanisms of gas permeation in the polymer matrix, such as 3D nanofillers and porous nanosheets. Differently from other reviews recently reported [23–27], a systematical assessment of the impact of different porous nanomaterials on the CO_2 separation performance of polymeric matrices is proposed, limiting the analysis mainly to the results reported in the last five years. The benefits that are related to the addition of the different porous nanofillers are discussed, categorizing the hybrid membranes according to the nature of the dispersed phases. The performances that are achieved by each dispersed phase are analyzed and compared among different polymeric matrices and loadings. This systematical analysis allows to identify the benefits and issues of each nanofiller type, offering an interesting tool to shape the direction of future research. The CO_2 separation performance are analyzed for the gas pairs of interest for carbon capture (CO_2 vs. N_2 and CO_2 vs. H_2) and for natural gas and biogas purification (CO_2 vs. CH_4). If no numerical values were reported in the original manuscript to describe the performance, relevant information were carefully extracted via plots' digitalization (WebPlotDigitizer, Version 4.1).

2. Metal Organic Frameworks (MOFs)

MOFs represent a heterogeneous class of hybrid materials constructed from organic bridging ligands and inorganic metal nods [28]. When compared to traditional porous materials, such as zeolites,

MOFs have drawn considerable attention thanks to their porous structure, large pore volume, fine tunable chemistry, and high surface area. MOFs are used in a large variety of applications, such as catalysis, sensing and electronic devices, drug delivery, energy storage, and gas separation [29–31]. In gas separation applications, recently, several efforts have been dedicated to the incorporation of MOFs in polymeric matrixes to produce hybrid membranes [20]. When compared to fully inorganic materials, such as Zeolites, the presence of organic ligands in the MOFs' structure leads to better affinity and adhesion with polymers and organic materials [6], making MOFs extremely promising for the achievement of proper interface morphology, and thus, improved separation performance. Hydrothermal, solvothermal or sonication-assisted methods, microwave-assisted, and room temperature reaction are the synthesis procedures that are frequently reported for MOFs [32]. Surface porosity, pore volume, and particle size of MOFs can be finely tuned by controlling the effective synthesis parameters, such as temperature, concentration, time, and pH. Theoretically, the unlimited number of ligands and metal ions provide infinite MOFs combinations.

MOFs frameworks can be either rigid or flexible. Rigid MOFs with tuned pore diameter could be a promising alternative to molecular sieves. The sieving behavior in rigid MOFs gives rise to considerably enhanced diffusion selectivity of gas pairs with different kinetic diameters, such as CO_2/N_2 or CO_2/CH_4. On the other hand, flexible structures undergo a considerable framework relaxation in the presence of external stimuli, such as host-gas interaction, pressure, temperature, or light [33–35]. Typically, this temporary structural transformability is a non-desirable effect, as it alters the initial sieving ability of the MOF structure [36]. The main structural rearrangements are typically referred as "gate opening" and "breathing" [33]. The former phenomenon is described as a transition from a closed and nonporous to a porous with open gates configuration upon the effect of external stimuli. As an example, ZIF-8 shows the swing in the imidazole linker and opening the narrow window at low to high pressure [37]. On the other side, the breathing effect is described as the abrupt expansion or compression of the unit cell. This is typically observed in MILs, where the structural transformation is referred as open pore, closed pore (cp), narrow pore (np), and large pore (lp) [34]. Linker rotation is another possible structural change, which is typically observed for UiO-66, where the benzene ring present on the organic ligand shows a rotational barrier that can be overcome at higher temperature [38,39]. Other important parameters that affect the transport properties of MOF nanoparticles are the pore volume and the surface area, as they mainly affect the gas sorption capacity of the MOF nanoparticles. In the case of CO_2, for example, it has been reported that the presence of unsaturated open metal sites can greatly enhance the CO_2 sorption capacity due to considerable polarizability and quadrupole moment. Open metal cations play as Lewis acidic nodes that strongly favors CO_2 [40,41]. The occurrence of breathing is reported to significantly affect the pore volume, and, therefore, the gas sorption ability. For example, in the case of MIL-53, an expansion of the unit cell volume from 1012.8 $Å^3$ to 1522.5 $Å^3$ when the CO_2 pressure is increased from 5 bar to 15 bar has been observed [36].

In the following sections, common MOFs that are used in fabricating mixed matrix membranes (MMMs) for CO_2 separation have been grouped according to their type of metal ion constituting the MOFs' architecture. Individual analyses of gas permeation have been dedicated to the MMMs containing Zeolitic Imidazolate Frameworks (translational metal ions), UiO-66 (Zr-based), CO_2-philic MOFs (Cu-based) and Materials Institute Lavoisier MOFs (trivalent metal ions). Other new and emerging MOFs have also been listed together in a separate section.

2.1. Zeolitic Imidazolate Frameworks (ZIFs)

Zeolitic imidazolate frameworks, known as ZIFs, have received great attention due to their exceptional transport properties [42]. Generally, ZIFs are a subclass of metal organic frameworks with a zeolite, like topology, consisting of large cavities linked by narrow apertures [1]. ZIFs are composed of M-Im-M, where M stands for transitional metal ions (such as Zn, Cr) and Im is the organic linker (imidazolate and its derivatives), respectively. M-Im-M forms a 145° angle, which is similar to

Si-O-Si angle in conventional aluminosilicate zeolites and makes structures analogous to zeolites with topologies of *sod*, *rho*, *gme*, *lta*, and *ana* [30,43]. Among the different ZIFs that are available [42], ZIF-7, ZIF-8, ZIF-11, ZIF-71, and ZIF-90 (Figure 1) are the most common MOFs incorporated in polymer matrix to produce hybrid membranes for carbon capture applications.

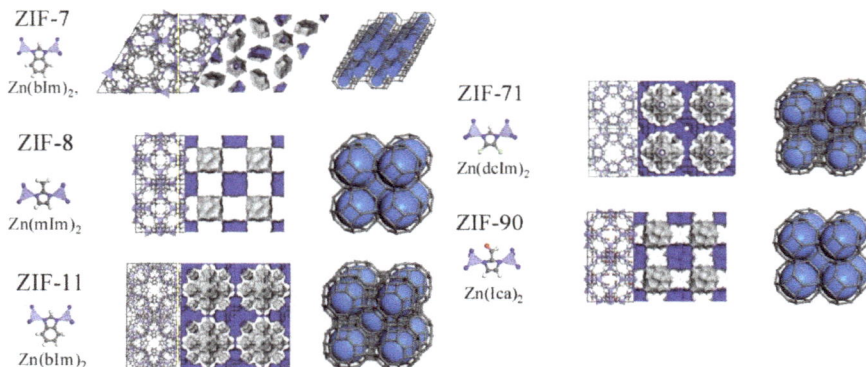

Figure 1. Zeolitic Imidazolate Frameworks (ZIF) structures with building blocks, topology, and accessible surface area for a probe diameter of 2 Å. Adapted from [42], with copyright permission from © 2012, Royal Society of Chemistry.

2.1.1. ZIF-8

ZIF-8 with *sod*-type topology and tetrahedral structure is the most frequently investigated MOF among the ZIFs family, which exhibits good thermal and exceptional chemical stability [44,45]. ZIF-8 has large pores of 11.8 Å and the pore limiting diameter of 3.4 Å, which represents a perfect sieving range for gas separation, such as CO_2/N_2 and CO_2/CH_4 [43]. However, the ZIF-8 framework is rather flexible, owing to the swing effect of organic linker that significantly affects the sieving ability [37,46]. This swing effect, which is supported experimentally and theoretically, was described by the rotation of imidazolate linker oscillating between two configurations of open window and close window [47]. The separation properties of ZIFs have been examined and researchers have explored their potential in the use of composite membranes for gas separation.

Matrimid® is a commercial glassy polyimide, which is widely used as polymer basis for comparison of MOFs' separation performance. Ordonez et al. [48] fabricated ZIF-8/Matrimid® mixed matrix membranes with nanoparticles loading up to 80 wt.% and investigated their transport properties for CO_2/N_2 and CO_2/CH_4 separation at 2.6 bar and 35 °C. ZIF-8 with a size range within 50–150 nm were dispersed in chloroform together with the polymer and self-standing membranes were obtained via solvent casting and dried at 240 °C under vacuum. While increasing the ZIF-8 loading, the tensile strength of the hybrid matrix dropped significantly and samples with 80 wt.% loading were found too brittle to be tested. Interestingly, the analysis of the transport properties showed a double behavior of the hybrids. Up to 40 wt.%, the disruption of the chain packing that is produced by the presence of the nanoparticles resulted in an increase in free volume, and consequently, in gas permeability. A 158% increase in CO_2 permeability (Table 1) was observed, even though the variation took place independent from the gas nature. On the contrary, at 50 and 60 wt.% loading the gas permeability dropped significantly, showing a considerable increase in the selective feature (CO_2/CH_4). The authors suggested a transition from a polymer-based to a ZIF-8-regulated transport, with the sieving effect of the fillers becoming dominant above a certain inorganic content. Interestingly, despite the CO_2-philic nature of ZIF-8, the hybrid samples maintained the H_2-selective features of the neat polymer (Table 1), but the low selectivity values ($H_2/CO_2 < 5$) are not of interest for the industrial applications. The CO_2 separation performances of ZIF-8/Matrimid hybrid membranes have also been investigated by Basu et al. [49], limiting the loading up to 30 wt.%. SEM imaging showed the formation of a proper interface morphology between the particles and the polymer phase. Similar to the previous case, the CO_2 permeability increased proportionally to the loading, reaching a 209% enhancement when compared to the neat polymer at the maximum loading. Possibly, the larger enhancement compared to the previous case may be attributed to the larger ZIF-8 size (250–500 nm). However, the separation factor appeared to be hardly affected by the presence of nanoparticles, with a maximum enhancement of 15%. Interestingly, the authors also compared the performance of other two MOFs (MIL-53 and $Cu_3(BTC)_2$), observing that the enhancement in CO_2 permeability is mainly dependent on the loading, whereas the nanoparticles nature and size play a minor role in affecting the transport properties. Song et al. [50] synthesized ZIF-8 with particle size of about 60 nm, and fabricated mixed matrix membranes by embedding them into Matrimid. Morphological analysis showed a proper polymer/particle interface up to the maximum loading investigated (30 wt.%). Notably, the smaller ZIF-8 size determined a 250% enhancement in CO_2 permeability at the highest loading, even though a negative effect on selectivity was observed (25% decrease at 30 wt.% loading) for both CO_2/N_2 and CO_2/CH_4.

Sonication has also been reported to be an important factor affecting the performance of ZIF-8-based mixed matrix membranes [51]. ZIF-8 nanoparticles were dispersed into Matrimid, exposing the casting solution to direct (sonication horn) or indirect (sonication bath) ultrasound wave (Figure 2). The study showed that different sonication intensities produced a significant change in the morphology of the nanoparticles, with limited influence on crystallinity and microporosity. When higher sonication intensity was applied to the casting solution, a proper interfacial morphology was achieved, with a simultaneous increase of permeability and selectivity (Table 1) and full consistency with the Maxwell model. When indirect sonication was employed, nanoparticles agglomeration was observed, affecting the efficiency of the hybrid membranes. ZIF-8 modification using mixed organic ligand (2-aminobenzimidazole as a substitution linker) has also been reported [52], leading to differences in pore size distribution and porosity when compared to pristine ZIF-8. When hybrid membranes were prepared while using Matrimid as polymer phase, no gate opening effect or structural flexibility was observed, and the ideal selectivity improved (Table 1). An interesting approach to improve the interface morphology has been proposed by Casado Coterillo et al. [53], who fabricated a ternary system, embedding ZIF-8 in a polymer matrix composed of Chitosan and [Emim][Ac]. At low ZIF-8 loading (5 wt.%), they achieved the best CO_2/N_2 separation performance and attributed the effect to a better adhesion between the Chitosan and the ZIF-8 phase that is offered by the presence of the ionic liquid at the interface.

Carter et al. [54] loaded 10% ZIF-8 with particle size of 95 nm in Matrimid and prepared two different dense membrane films with aggregated ZIF-8 nanoparticles and with a homogeneous dispersion. As expected, the single gas permeation tests showed improved selectivity and permeability for the well-dispersed membrane and the lower drop observed for the N_2 permeability, with respect to CH_4 permeability, was explained in terms of surface diffusion mechanism and framework flexibility of ZIF-8. Again, the addition of ZIF-8 nanoparticles enhanced the H_2-selective properties of the hybrids, with the aggregated samples showing even better performance (68% increase in H_2 permeability) when compared to the one with homogeneous dispersion (Table 1). However, the selectivity remained too low ($H_2/CO_2 < 5$) to become valuable for real H_2 purification. Interestingly, the reported analysis of hybrid membranes based on Matrimid and ZIF-8 clearly showed that synthesis protocol, particle size, and possible modification play a major role in the determination of the membrane performance. Guo et al. [55] recently investigated the effect of ZIF-8 nanoparticles on another commercial polyimide, P84. As reported for Matrimid, the CO_2 permeability increased proportionally to the MOF content. Also, the CO_2/CH_4 selectivity increased remarkably, but at the highest loading (31 wt.%), a drop (Table 1) was observed. A drop in the diffusion selectivity was measured (Figure 3), clearly suggesting that the formation of interfacial voids that are associated to MOFs aggregation is responsible for the observed phenomenon.

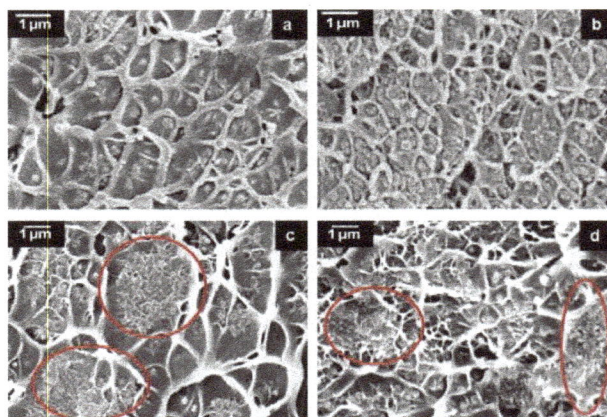

Figure 2. Dispersion of ZIF-8 by direct (**a,b**) and indirect (**c,d**) sonication of 10 wt.% (**a,c**) and 25 wt.% (**b,d**) loading in Matrimid [51], with copyright permission from © 2012 Elsevier.

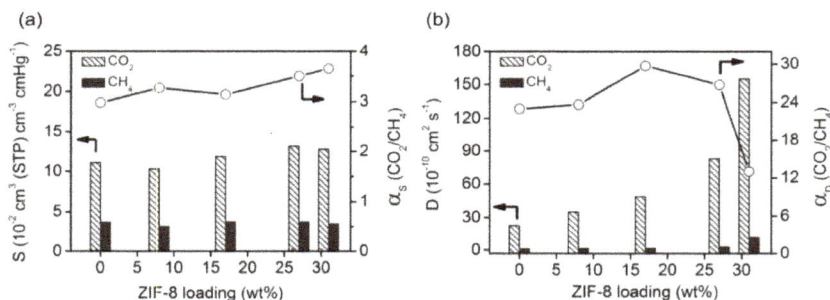

Figure 3. Effect of ZIF-8 loading on the solubility and the diffusivity selectivity when embedded in P84 polyimide [55], with copyright permission from © 2018 Elsevier.

6FDA is another glassy polyimide that has been largely investigated for the fabrication of ZIF-based mixed matrix membranes. The higher free volume when compared to Matrimid allows for the 6FDA polymer family to achieve larger gas permeation, offering a more suitable option for industrial applications. Jusoh et al. [56] reported significant improvement in CO_2 permeability of 6FDA-durene by embedding up to 20 wt.% ZIF-8 in the polymer matrix. An optimum loading of 10 wt.% was identified (Table 1), as a further increase of the inorganic content led to negligible enhancement of CO_2 permeability, but a significant decrease of CO_2/CH_4 selectivity. Furthermore, the gas separation enhancement of ZIF-8/6FDA-durene was attributed to the influence of pore limiting diameter and quadrupole interaction of CO_2 with the ligand in ZIF-8 framework. Wijenayake et al. [57] proposed surface crosslinking as possible approach to improve the performance of 6FDA-based hybrid membranes containing ZIF-8 nanoparticles. The addition of 33 wt.% ZIF-8 in the polymer matrix enhanced significantly the CO_2 permeability (~400%, Table 1), reaching up to ~1500 Barrer, similar to the one that was observed in the previous study. The effect on the selectivity was limited. Even though post-synthetic modification of ZIF-8 using ethylenediamine showed enhanced CO_2 adsorption capacity [58], the use of ethylenediamine vapors to crosslink the surface of the hybrid membrane led to a limited improvement on the CO_2 selectivity along with a drastic drop in CO_2 permeability. As in the case of Matrimid, the addition of ZIF-8 to 6FDA polyimide improved the H_2-selective feature, and a H_2/CO_2 selectivity of 12 has been achieved upon surface modification. Askari and Chung [59] studied the effect of annealing temperature on the performance of 20 wt.% ZIF-8 containing 6FDA-durene mixed matrix membrane by heating to different temperature (200, 350, and 400 °C) below glass transition temperature ($T_g > 400$ °C). The highest gas permeability was obtained for 20 wt.% loaded membrane annealed at 400 °C (from 487 Barrer at 200 °C to 1090 Barrer at 400 °C) and the contribution of the inorganic phase was enhanced at higher annealing temperatures. When the cross-linkable co-polyimide (6FDA-durene/DABA) was used in the place of the homopolymer, higher selectivity values could be achieved, but the improvement took place to the detriment of CO_2 permeability. Nafisi and Hägg investigated the gas separation performance of ZIF-8 containing membrane prepared using 6FDA-durene [60] and PEBAX 2533 [61] (a commercial polyether-block-amide) as polymer phase. In both cases, the CO_2 permeability increased along with the inorganic content, but the influence of ZIF-8 nanoparticles appeared to be more effective for PEBAX 2533. At 30 wt.% loading, a 50% enhancement of CO_2 permeability (2186 Barrer) was observed for 6FDA-durene whereas a ZIF-8 loading of 35 wt.% in PEBAX 2533 corresponded to a 3.6-fold improvement of the CO_2 permeability (1287 Barrer). Furthermore, at high inorganic loading, the polyimide showed reduced CO_2 selectivity, whereas negligible effect on the separation performance was observed for PEBAX.

Recently, Sanchez-Lainez et al. [62] reported the fabrication of mixed matrix membranes based on polybenzimidazole (PBI), obtained via phase inversion method for H_2/CO_2 separation. At 180 °C, the presence of the ZIF-8 nanoparticles improved the H_2/CO_2 selectivity as well as the H_2 permeance. At higher temperature (250 °C), the presence of defects resulted in a drop in the selective characteristic, but higher feed pressure (3 bar vs 6 bar) restored the H_2/CO_2 selectivity to a value close to 20.

Recent publications showed an increasing research also on the fabrication of thin composite membranes containing ZIF-8 nanoparticles. Dai et al. [63] fabricated asymmetric hollow fiber mixed matrix membranes using dry jet-wet quench method. In particular, they dispersed 13 wt.% ZIF-8 nanoparticles (size ~200 nm) into a polyetherimide (Ultem 1000) matrix. CO_2/N_2 separation performance for the HF membranes were tested at 35 °C and 100 psi. For both pure and mixed gas, the separation performance was improved. The permeance and selectivity of the ZIF-8 containing hollow fibers improved by 85% and 20%, respectively, when compared to the unloaded hollow fibers. Higher selective feature were observed for mixed gas conditions using 20 vol.% CO_2 in the feed. A comprehensive review on progresses and trends on hollow fiber mixed matrix membranes has been recently reported by Mubashir et al. [64]. The review includes a comparison between the results obtained for flat sheet and hollow fiber mixed matrix membranes at similar filler loading and operating conditions. It was concluded that hollow fiber mixed matrix membranes that are loaded with ZIF-8, ZIF-93, and amine functionalized MILs show higher separation performance for CO_2/N_2 and CO_2/CH_4.

Thin film can be obtained also by coating on porous support. Thin film composite membranes and thin film nanocomposite membrane containing MOFs have been developed for nanofiltration and organic solvent separation [65–68]. However, only few studies can be found in literature investigating the gas transport properties of thin hybrid selective layers. Sánchez-Laínez et al. [69] reported a novel ultra-permeable thin film nanocomposite (TFN) containing ZIF-8 for H_2/CO_2 separation. The selective layer (50–100 nm) was formed on a polyimide P84 asymmetric support. The nanoparticles were dispersed in different loadings (0.2, 0.4, and 0.8% w/v) in a polyamide matrix. The incorporation of ZIF-8 nanoparticles enhanced the gas separation performance. At 35 °C and 0.4% w/v ZIF content, a 3-fold increase in selectivity was observed compared to the pristine polymer. An increase in the temperature had a positive impact on the performance, especially in terms of H_2 permeance (up to 988 GPU at 250 °C for the pristine polymer). At 180 °C, TFN membranes containing 0.2 and 0.4% (w/v) of ZIF-8 exhibited a marked selectivity increase of 42% and 64%, respectively. At higher loading

(0.8% w/v), the presence of micro voids and defects determined a significant drop in both permeance and selectivity. A further increase in temperature led to higher H_2 permeance of TFN membranes with negligible influence on the selective features.

2.1.2. ZIF-7

ZIF-7 is another promising candidate of the ZIFs family for gas separation applications. 1H-benzimidazole is the bridging ligand, which is connected to the Zn metal clusters and creates a 3D sodalite topological framework (Figure 2). Its pore diameter ranges between 3 and 4.3 Å [44,70]. The narrow pore size makes ZIF-7 suitable for H_2 purification from CO_2. Nevertheless, due to the flexibility of the benzimidazole linker, ZIF-7 also shows the "gate opening effect", undergoing a reversible transition of the pores (from narrow to large framework flexibility of ZIF-7 that allows for gas molecules with a molecular diameter as large as 5.2 Å to access the pores and cavities). This gate opening effect of ZIF-7 was observed in adsorption isotherms (CO_2, ethane, and ethylene) [71].

Li et al. [72] evaluated the separation performance of ultrathin hybrid membrane composed by a poly(amide-b-ethylene oxide) (Pebax 1657) and ZIF-7 nanoparticles. ZIF-7 particles with a size between 40 and 50 nm were synthesized and embedded up to 34 wt.% within the polymer matrix. Subsequently, thin composite membranes were prepared by coating the casting solution on a porous PAN support (PTMSP gutter layer was used to prevent pore penetration of the selective layer). Increasing the ZIF-7 loading up to 22 wt.% showed a remarkable increase (Table 1) in both CO_2 permeability and CO_2/CH_4 and CO_2/N_2 ideal selectivity. However, at higher loading (34 wt.%) polymer rigidification around the nanoparticles took place, positively affecting the selectivity (214% and 208% enhancement for CO_2/CH_4 and CO_2/N_2, respectively), while the CO_2 permeability was considerably lower when compared to that of the neat polymer. Post synthesis modification of nanosized (40–70 nm) ZIF-7 was implemented by Al-Maythalony et al. [73], aiming at tuning the pore size by exchanging the organic ligand, benzimidazolate with benzotriazolate. The synthesized nZIF-7 and PSM-nZIF-7 were embedded in a polyetherimide (PEI) matrix. The post synthesis modification resulted in an increase of CO_2 permeability of all the examined gases (N_2, CH_4, and CO_2 by 737%, 470%, and 198%, respectively). Nevertheless, the bigger enhancement of gases with larger kinetic diameters reduced the CO_2-selective feature of the hybrids when compared to the pristine PEI.

2.1.3. ZIF-11, ZIF-71, and ZIF-90

ZIF-11, ZIF-71, and ZIF-90 are the other three structures from the ZIFs library that are of interest for gas separation applications and are characterized by *rho* (for both ZIF-11 and ZIF-71), and *sod* type topology with apertures of 3 Å, 4.2 Å, and 3.5 Å, respectively [45,74]. ZIF-90 is an attractive MOF for CO_2 capture owing to its covalent carbonyl bond in the imidazole linker favoring CO_2 and the 0.35 nm of pore size, which is suitable for CO_2/CH_4 separation. Alternatively, ZIF-71 is selected due to its large cavity pore diameter (16.5 Å) when compared to that of ZIF-8, ZIF-90, and ZIF-11 (cavity pore diameter 11.6, 11.2, and 14.6 Å, respectively) that has the potential to enhance the gas separation performance of hybrid membranes [75,76].

Ehsani and Pakizeh [77] examined the performance of hybrid membranes with a ZIF-11 loading range of 10–70 wt.% incorporated into PEBAX 2533. Morphological characterization of MMMs revealed an excellent adhesion between the polymer matrix and the nanoparticles. Even at 50 to 70 wt.% ZIF-11 loading, no significant agglomeration could be observed, even though poorer interfacial morphology appeared. At lower MOF loading, the presence of polymer chain rigidification and pore blockage resulted in a gas permeability reduction (~20%). At higher loading (>50 wt.%), the CO_2 permeability increased when compared to pristine polymeric membrane, reaching a value of 403 Barrer at 70 wt.% (Table 1). Different effects were observed for selectivity: the CO_2/CH_4 selectivity increased from 8 to 12.5 at increasing the MOF content, but a negative trend was observed in the case of CO_2/N_2 selectivity. ZIF-11 has also been embedded in 6FDA-DAM polyimide [78]. SEM micrographs showed no apparent agglomeration for loading up to 30 wt.%. An optimum was observed incorporating

20 wt.% ZIF-11, leading to a 12-fold enhancement of CO_2 permeability (Table 1), with limited effect on the ideal selectivity. The CO_2 permeability improvement was associated to the achievement of particles alignment, and subsequently, an increase in fractional free volume of the hybrid matrix, which is confirmed by d-spacing analysis. The lack of selectivity improvement for 20 wt.% loading was related to the much higher gas permeability of ZIF-11 as compared to 6FDA-DAM, as predicted by the Maxwell model. Further increase in ZIF-11 loading did not show any improvement of the separation performance, owing to polymer chain rigidification and pore blockage.

Hybrid membranes based on PIM-1 and ZIF-71 with various loading were fabricated by Hao et al. [79]. The addition of ZIF-71 into PIM-1 considerably improved the gas transport, and in the case of CO_2, the permeability value increased from 3295 to 8377 Barrer (Table 1). Photo oxidation obtained via UV treatment of the neat polymeric matrix increased the ideal selectivity to the detriment of gas permeability. As expected, the presence of the nanofillers helped in minimizing the gas permeability drop, showing impressive membrane performance (CO_2 permeability of 3459 Barrer, CO_2/CH_4 and CO_2/N_2 selectivity of 35.6 and 26.9, respectively) [79]. The effect of particle size (30, 200, and 600 nm, as seen in Figure 4) has also been investigated, using a fixed amount of nanoparticles in 6FDA-durene (Table 1) [76]. The permeability enhancement associated to the presence of the nanoparticles did not scale with the particle size, but it showed an optimum when the 200 nm particles size were used. In addition, the negligible effect on the ideal gas selectivity suggested the existence of a trade-off between the particle size and the gas separation performance, giving an important indication for the further development of nano-hybrid membranes.

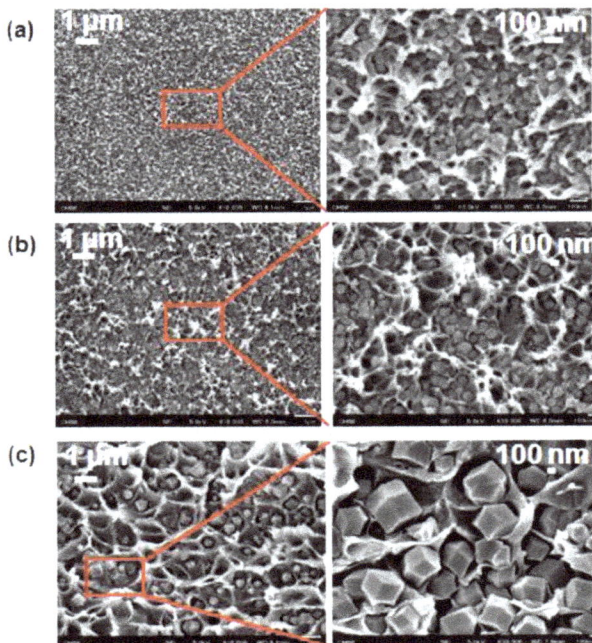

Figure 4. Cross-sectional morphology of 6FDA-Durene containing ZIF71 particles with average size of 30 nm (**a**); 200 nm (**b**); and 600 nm (**c**) [76], with copyright permission from © 2016, American Chemical Society.

Bae et al. [80] studied the CO_2 separation performance of MMMs containing a fixed amount of ZIF-90 (15 wt.%), coupled with three different polyimides (6FDA-DAM, Matrimid and Ultem), aiming at determining the effect of the nanofillers on different polymer phases. In the case of Matrimid and Ultem, the CO_2 permeability increased (~100%, Table 1). As previously reported, the negligible selectivity variation observed is related to the higher gas permeability of the nanoparticles, as predicted by the Maxwell model. When a more permeable matrix was used (6FDA-DAM), the CO_2 permeability improvement was followed by an increase of the CO_2-selective features of the hybrid matrix. Mixed gas permeation tests showed separation performances well above the CO_2/CH_4 and CO_2/N_2 upper bounds. Zhang et al. [81] utilized ZIF-90 as the filler in triptycene-based polymer and prepared hybrid membranes for CO_2/N_2 and CO_2/CH_4 separation. Cross-sectional SEM images revealed homogenous dispersion of the nanofillers and membranes with defect-free interfacial morphology, even at high loadings. The membrane containing 50 wt.% ZIF-90 showed a 215% increase of the CO_2 permeability value (Table 1), without sacrificing the gas selectivity. The ability of ZIF-90 to disrupt the polymer chain packing, and consequently, increase in free volume, was also suggested as source of additional permeability enhancement.

Table 1. Gas separation performance of ZIFs-b ased mixed matrix membranes (operating conditions ranging within 1–5 bar, 20–35 °C, unless differently specified).

Filler	Polymer	Loading (wt.%)	P_{CO_2} (Barrer)	α_{CO_2/N_2}	α_{CO_2/CH_4}	α_{CO_2/H_2}	Ref.
			ZIF-8				
ZIF-8 50–150 nm	Matrimid 5218	0	9.5	30.7	39.8	0.34	[48]
		20	9.0	30.1	51.1	0.29	
		30	14.2	24.1	38.2	0.31	
		40	24.5	23.4	27.8	0.35	
		50	4.7	26.2	124.9	0.35	
		60	8.1	18.4	80.7	0.26	
ZIF-8 250–500 nm	Matrimid 9725	0	0.21 [a]		28.0		[49]
		10	0.31 [a]		29.5		
		20	0.42 [a]		31.0		
		30	0.7 [a]		31.5		
ZIF-8 60 nm	Matrimid 5218	0	8.1	22.4	35.2		[50]
		5	10.1	21.2	39.1		
		10	13.7	21.6	30.6		
		20	16.6	19.0	35.8		
		30	28.7	17.1	24.9		
ZIF-8 Dir. Son.	Matrimid	0	10.7		33.9		[51]
		10	21.9		36.0		
		25	47.0		39.0		
Indir. Son.		10	13.2		31.0		
		25	23.2		31.9		
ZIF-8 ZIF-8-ambz	Matrimid 5218	0	9.0		35.0		[52]
		15	11.3		35.0		
		15	10.4		36.5		
		30	10.2		38.0		
ZIF-8 95 nm	Matrimid 5218	0	9.5	13.6	29.8	0.31	[54]
		10	13.1	20.5		0.26	
		10 [b]	15.5	26.7	34.4	0.34	
ZIF-8 30 nm	P84	0	2.7 [c]		54.1		[55]
		8	3.2 [c]		63.5		
		17	6.3 [c]		93.6		
		27	11.0 [c]		92.3		
		31	20.0 [c]		45.8		

Table 1. *Cont.*

Filler	Polymer	Loading (wt.%)	P_{CO_2} (Barrer)	α_{CO_2/N_2}	α_{CO_2/CH_4}	α_{CO_2/H_2}	Ref.
ZIF-8 50 nm	6FDA-durene	0	468		7		[56]
		5	694		16.5		
		10	1427		28.7		
		15	1466		11.3		
		20	1463		8.97		
ZIF-8	6FDA-durene	0	469	13.4	15.6	0.91	[57]
		33	1553	11.3	11.1	0.71	
		33 [d]	23.7	11.8	16.9	0.08	
ZIF-8 80 nm	6FDA-durene T = 200 °C	0	352		16.6		[59]
		20	487		17.9		
	T = 350 °C	0	432		13.8		
		20	857		13.1		
	T = 400 °C	0	541		13.1		
		20	1090		13.0		
ZIF-8 100–200 nm	6FDA-durene	0	1468	25.4	22.6		[60]
		3	1593	25.7	21.9		
		5	1695	22.7	20.1		
		7	1774	22.1	19.4		
		10	1882	20.5	19		
		15	1940	18.6	18.1		
		20	2027	17.5	16.9		
		30	2186	17	17.1		
ZIF-8	PEBAX 2533	0	351	35.1	8.3		[61]
		5	305	25.4	6.8		
		10	427	30.5	8.5		
		15	574	30.2	10.4		
		20	854	28.5	9.2		
		25	1082	30.9	8.5		
		30	1176	31.8	8.7		
		35	1287	32.2	9		
ZIF-8	Ultem 1000	0	14 [e]	30			[63]
		13	26 [e]	36			
ZIF-7							
ZIF-7 40–50 nm	PEBAX 1657	0	72	34	14		[72]
		8	145	68	23		
		22	111	97	30		
		34	41	105	44		
ZIF-7 PSM-ZIF-7 [g]	PEI	0	82.5	3.8	4.4		[73]
		5	64.7	17	12.9		
		5	246	1.3	2.3		
ZIF-11							
ZIF-11 500–5000 nm	PEBAX 2533	0	232	41.3	8		[77]
		10	212	53	9.7		
		30	186	47.9	11.4		
		50	233	46.9	11.2		
		70	402	29	12.4		
ZIF-11 200–2000 nm	6FDA-DAM	0	21.4		32.7		[78]
		10	107		31.3		
		20	273		31		
		30	76.7		30.4		

Table 1. *Cont.*

Filler	Polymer	Loading (wt.%)	P_{CO_2} (Barrer)	α_{CO_2/N_2}	α_{CO_2/CH_4}	α_{CO_2/H_2}	Ref.
			ZIF-71				
	PIM-1	0	3265	20.1	10.2		[79]
ZIF-71		10	4271	19.4	11.3		
<1000 nm		20	5942	20	11.9		
		30	8377	18.3	11.2		
	UV-PIM-1	0	1233	29.8	34.1		
UV-ZIF-71		10	1909	29.1	35.5		
<1000 nm		20	2546	27.2	35.3		
		30	3459	26.9	35.6		
ZIF-71	6FDA-Durene	0	805	14.7	17		[76]
30 nm		20	2560	13.8	14.2		
200 nm		20	2744	13.2	13.9		
600 nm		20	1656	13.5	14.7		
			ZIF-90				
	6FDA-DAM	0	402		17.5		[80]
ZIF-90		15	808		27.2		
810 nm	Ultem®1000	0	1.4		37.9		
ZIF-90		15	2.9		38.9		
	Matrimid	0	7.7		34.9		
ZIF-90		15	12.1		34.8		
	6FDA-DAM [h]	0	390		24		
		15	720		37		
	6FDA-TP [i]	0	20	20	37		[81]
ZIF-90		10	26	24	42		
60–105 nm		20	29	22	38		
		40	45	20	36		
		50	63	20	36		

[a] Permeance (GPU), membrane thickness 40–65 μm; [b] ZIF-8 synthesized using the solution collected from freshly-synthesized ZIF-8 dope after centrifugation; [c] equimolar CO_2/CH_4 mixture; [d] membrane surface cross-linked using ethylenediamine vapour; [e] Permeance (GPU), membrane thickness ~60 μm; [f] Permeance (GPU), membrane thickness 50–100 nm; [g] PSM: post-synthetic modification; [h] gaseous mixture as feed gas; [i] TP: triptycene, 10 atm feed pressure.

According to the analysis of different ZIFs in different polymeric materials, it appears that it is possible to achieve relatively high loading of isotropic ZIFs particles in the polymer matrix (up to 60 wt.%). However, the optimum concentration of inorganic nanofillers appeared to be in the range of 30 to 40 wt.%; at higher loading, no significant benefits for CO_2 permeability can be obtained, but a decrease in selectivity can be expected. The use of ZIFs has been demonstrated to also be successful for highly permeable polymer (6FDA-based polyimides, PIM-1, PEBAX), and typically the introduction of nanoparticles has the main function of disrupting the polymer chain packing and increasing the free volume in the hybrid matrix. However, despite the achievement of suitable interface morphology, the addition of ZIFs to polymer matrix seldom is reported to have a significant impact on the selective feature of the mixed matrix membrane. Among the investigated ZIFs, it is not possible to identify one type that is able to stand out, but the efficiency of each type also depends on the chosen polymeric phase and the synthetic procedures. Furthermore, ZIF nanoparticles with smaller size appears to be more effective when compared to inorganic phases with bigger average size. Finally, despite the CO_2-philic nature of the nanofillers, the incorporation of ZIFs in polymeric matrix typically enhances the H_2-selective feature of the pristine polymeric matrix.

2.2. Zirconium 1,4-Dicarboxybenzene (UiO-66)

UiO-66 is a zirconium-based metal-organic framework that is built from zirconium oxide ($Zr_6O_4(OH)_4$) nodes linked together by 1,4-benzendicarboxylate as a bridging ligand [82]. UiO-66 is the first member of zirconium based MOFs family with *fcu*-topology introduced by Cavka et al. [83]. It owns a Langmuir surface area of 1187 m^2/g and the narrow triangular windows that are present in the UiO-66 framework have different sizes (Figure 5): 6 Å is the window connected to the two octahedral cages, with the size of 11 Å, and the tetrahedral cage, which has an opening of 8 Å (Figure 5). UiO-66 showed exceptional mechanical and chemical stability on exposure to high temperature, up to 500 °C, and chemicals, making this MOF a promising candidate for many applications [82,84]. The benzene ring has been found to be characterized by the rotational barrier as compared to other MOFs, leading to changes in the pore opening size (Figure 5C), and this effect showed a temperature dependency behavior [85].

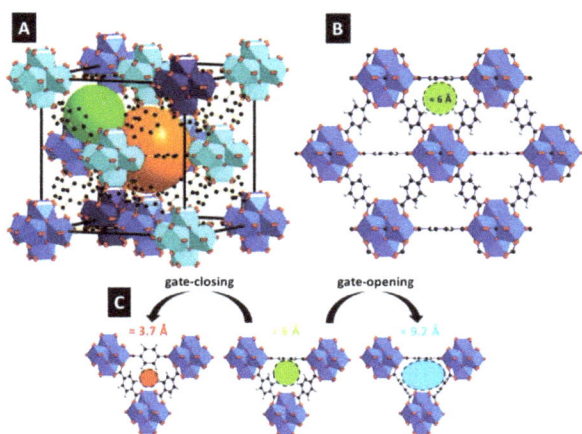

Figure 5. Three-dimensional (3D) structure of UiO-66 (**A**) visualizing the octahedral cage (orange) and the tetrahedral cage (green). Triangular windows (**B**) between the octahedral and tetrahedral cages. Pore opening changes upon rotation of the benzene ligands (**C**) [84], with copyright permission from © 2017, American Chemical Society.

Hybrid membranes embedding 5 to 20 wt.% pristine UiO-66 and amine functionalized UiO-66-NH_2 (average size 60–80 nm) in PEBAX 1657 have been prepared [86]. For both types of nanoparticles, the CO_2 permeability increased proportionally to the amount of inorganic phase, reaching a ~2.5-fold enhancement of the pristine polymer value (Table 2). These results suggested that UiO-66 showed a strong affinity towards CO_2 due to the presence of OH coordinated bond connected to Zr cluster. Different trends were observed for the CO_2/N_2 selectivity, which showed an optimum between 7.5 and 10 wt.% loading. The better affinity of the UiO-66-NH_2 with the polymer phase allowed for reaching better selectivity improvement (88%) as compared to the pristine MOF (42%). Interestingly, the mixed matrix membranes prepared with UiO-66-NH_2, retained stable performances even in the presence of humidity. Similar nanoparticles (UiO-66 and UiO-66-NH_2) have been embedded also into Matrimid 9725 [87]. The use of two modulators (benzoic acid, BA, and 4-aminobenzoic acid, ABA) was reported to allow for the linkage of the amine groups in different positions in the UiO-66 structure. The presence of the ABA modulator increased the CO_2/CH_4 selectivity up to 55% (from 31.2 to 47.4), together with a six-folds improvement of the CO_2 permeability for the amine-modified UiO-66 (Table 2). Surface modification of the UiO-66 has also been proposed as a possible approach to improve the nanoscale morphology at the organic/inorganic interface [88].

The surface modification was performed using phenyl acetyl (PA), decanoyl acetyl (DA), and succinic acid (SA) in order to enhance the interaction between nanoparticles and Matrimid 5218 was used as polymer phase. A good adhesion and interaction between surface functionalized UiO-66-NH$_2$ and polymer matrix was observed, leading to improved mechanical and chemical properties of and the hybrid membranes. 23 wt.% loading of PA-modified UiO-66-NH$_2$ enhanced the CO$_2$ permeability by 229% (from 8.5 Barrer to 28 Barrer), with a simultaneous improvement of CO$_2$/N$_2$ selectivity by 25%. The permeability and selectivity increased due to the strong interactions between the CO$_2$ and the NH$_2$ groups that are present in the MOF, together with interaction of imide group in Matrimid and aromatic ring in PA through π-π bonds. The poor interaction between fillers and Matrimid in DA and SA-modified UiO-66-NH$_2$ resulted in a reduction in permeability and selectivity when compared to PA-modified UiO-66-NH$_2$ particles.

The influence of amino and carboxylic group functionalization of UiO-66 have been investigated using PIM-1 as polymer phase [89]. The investigation considered "as-cast" and "solvent exchanged" PIM-1 membrane: the latter showed higher CO$_2$ permeability (8210 Barrer) compared to the pristine membrane (4770 Barrer), and the difference is attributed to the excess free volume that is generated by the solvent removal. The addition of pristine UiO-66 to the matrix generated an enhancement in CO$_2$ permeability (up to 59% for the "as cast" membrane and 32% for the "solvent exchanged" sample, Table 2) when compared to the pristine polymeric sample. In the case of UiO-66-NH$_2$ and UiO-66-(COOH)$_2$, the CO$_2$ permeability also showed an increase, but with a lower extent as compared to the pristine nanoparticles. In the case of the selectivity, the parameter showed a limited variation for both CO$_2$/N$_2$ (decrease up to 10%) and CO$_2$/CH$_4$ (decrease up to 20%) upon the addition of the nanoparticles, both pristine and functionalized. Performance for CO$_2$/H$_2$ separation were also reported. The pristine polymer showed a CO$_2$-philic behavior, which was slightly enhanced in presence of the nanoparticles (particularly in the case of UiO-66-(COOH)$_2$). However, the selectivity value remains too low to be attractive for industrial separations.

In another study [90], water modulation was employed to reduce the particle size of UiO-66 (from 100–200 to around 20–30 nm) and the water modulated nanoparticles (UiO-66-H) were further surface-modified using amine (UiO-66-NH$_2$) and bromide (UiO-66-Br) functional groups. The reduction in particle size improved the dispersion of UiO-66 into polymer matrix by minimizing the formation of non-selective microvoids. The decrease in the CO$_2$ selective feature of the hybrids observed with increasing the content of unmodified UiO-66 was therefore prevented (Figure 6), and a selectivity enhancement was observed for all of the modified nanoparticles (up to 71% and 95% in the case of CO$_2$/N$_2$ and CO$_2$/CH$_4$ selectivity for 10 wt.% UiO-66-NH$_2$ loading). This effect was mainly associated to the increased rigidity of interphase. However, the improved interactions between the functionalized nanoparticles and the polymer chains led to a negligible effect on CO$_2$ permeability, which instead was significantly enhanced (~100%) in the case of unmodified UiO-66. Despite the differences that were observed with respect to the previous study in terms of pristine PIM-1 transport properties, the performance achieved by embedding UiO-66 and UiO-66-NH$_2$ are similar, supporting the consistency of the results.

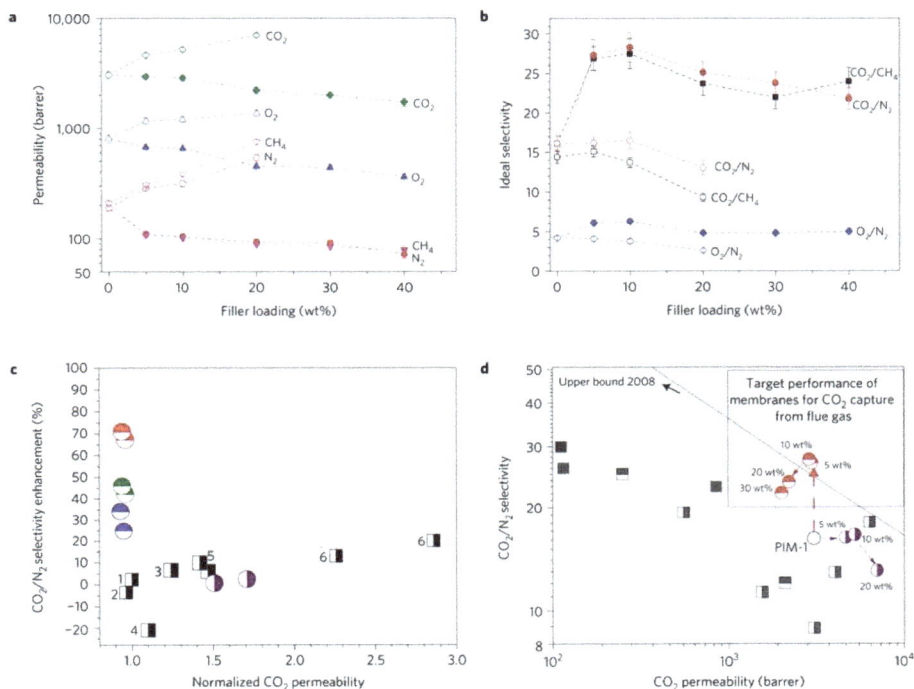

Figure 6. Gas permeability (**a**) and gas selectivity (**b**) of UiO-66-NH$_2$ (filled symbols) and pristine UiO-66 (open symbols) embedded in PIM-1. Comparison with literature results (**c**) and Robeson plot (**d**) for CO$_2$/N$_2$ separation [90], with copyright permission from © 2017, Springer Nature.

The influence of UiO-66 on the gas separation performance of 6FDA-based polyimides were also evaluated for mixed gas feed CO$_2$/CH$_4$ (50/50 v/v) [91]. Different 6FDA-based polymers were investigated (6FDA-BisP, 6FDA-ODA, and 6FDA-DAM). CO$_2$ permeability was found to increase proportionally to the inorganic content for all of the different polymer phases, even though a larger enhancement was observed for the low permeable ones. In the case of 6FDA-Bisp and 6FDA-ODA, CO$_2$ permeability improved by 357% and 178%, whereas for 6FDA-DAM, the enhancement was limited to 136%. The permeability improvement was associated to a FFV increase upon the incorporation of the inorganic phase, and the benefits was more pronounced for the polymer phase with an initially lower FFV. Improvement in terms of selectivity was observed for 6FDA-BisP and 6FDA-ODA up to 17 wt.% loading, but at higher loadings, poor nanoparticles dispersion determined a drop in the selective feature of the hybrids. Interestingly, a negligible effect was observed for the more permeable 6FDA-DAM. The authors also investigated the effect of surface functionalization of UiO-66 when embedded in 6FDA-DAM [92]. The amino-functionalized UiO-66-NH$_2$ was prepared via the direct synthesis method, and UiO-66-NH-COCH$_3$ was synthesized via post-synthetic modification of UiO-66-NH$_2$ using acetamide-ligand. When compared to the results that were obtained with the pristine UiO-66, the surface modification helped in achieving a better polymer-MOF interface, reducing the free volume of the hybrid matrix at a given loading. At low pressure, negligible effects were observed on the transport properties when the modified MOFs were used, but at higher feed pressure, the post-synthetic modification showed better results in terms of CO$_2$/CH$_4$ selectivity.

Table 2. Gas separation performance of UiO-66-based mixed matrix membranes (operating conditions ranging within 1–5 bar, 20–35 °C, unless differently specified).

Filler	Polymer	Loading (wt.%)	P_{CO_2} (Barrer)	α CO2/N2	α CO2/CH4	α CO2/H2	Ref.
	PEBAX 1657	0	51.5	42.1			[86]
UiO-66		5	75.0	56.0			
60–80 nm		7.5	90.0	60.0			
		10	96.3	56.6			
		12.5	110.5	40.0			
		15	115.0	27.0			
		20	134.0	21.0			
UiO-66-NH$_2$		5	71.0	68.0			
60–80 nm		7.5	78.0	76.0			
		10	87.0	79.2			
		12.5	96.0	45.0			
		15	100.0	37.5			
		20	122	26			
	Matrimid 9725 [a]	0	5.9		31.2		[87]
UiO-66		30	15.0		35.8		
UiO-66-BA		30	17.8		42.9		
UiO-66-ABA		30	13.6		45.1		
UiO-66-NH$_2$		30	17.8		37.3		
UiO-66-NH$_2$-BA		30	17.4		39.3		
UiO-66-NH$_2$-ABA		30	38.0		47.4		
	Matrimid 5218 [b]	0	8.5	29			[88]
UiO-66 -NH$_2$		12	18.5	33			
200 nm		23	24	36			
		40	28	27.5			
UiO-66-NH$_2$-PA		12	20.5	32.5			
		23	28	36.5			
		40	31	28			
UiO-66-NH$_2$-C10		23	22.5	28			
UiO-66-NH$_2$-SA		23	20	30.5			
	PIM-1	0	4770	21.8	16.7	1.76	[89]
UiO-66	as cast	9.1	5940	23.2	16.	1.93	
200 nm		16.6	7610	20.7	14.4	1.67	
		23.1	7610	20.7	14.4	1.67	
		28.6	4940	13.6	11.2	0.66	
UiO-66-(COOH)$_2$		9.1	4600	20.9	14.1	2.22	
200 nm		16.6	5190	20.4	13.2	2.19	
		23.1	5300	19.9	12.9	2.22	
		28.6	6090	20.6	15.2	1.63	
UiO-66-NH$_2$		9.1	4810	22.2	16.5	1.62	
200 nm		16.6	6340	20.9	14.9	2.03	
		23.1	5070	20.1	14.7	1.58	
		28.6	6310	21.5	13.3	2.10	
	PIM-1	0	8210	21.2	15.7	1.63	
UiO-66	exchanged solvent	16.6	9980	21.6	17	1.23	
200 nm		23.1	9980	21.6	17	1.23	
		28.6	10,900	15.2	13.2	1.74	
UiO-66-(COOH)$_2$		16.6	9720	18.9	11.7	2.28	
200 nm		23.1	8770	18.1	11	2.05	
		28.6	9020	22.1	13.5	1.02	
UiO-66-NH$_2$		9.1	8740	22	14.7	1.84	
200 nm		16.6	10,700	21.4	13.7	1.88	
		23.1	9570	23.4	13.8	1.43	
		28.6	9030	19.5	13	1.70	

Table 2. *Cont.*

Filler	Polymer	Loading (wt.%)	P_{CO_2} (Barrer)	α $_{CO2/N2}$	α $_{CO2/CH4}$	α $_{CO2/H2}$	Ref.
	PIM-1	0	3054	16.1	14.5	1.67	[90]
UiO-66		5	4620	16.2	15.1	1.90	
100–200 nm		10	5210	16.5	13.7	2.04	
		20	6981	13	9.3	2.60	
UiO-66-H		5	2765	22.9	18.2	0.88	
20–30 nm		10	2631	23.5	18.8	0.88	
		20	2606	24.6	20.1	0.89	
		30	1880	18.3	16.1	1.55	
		40	1023	21.4	15.8	1.67	
UiO-66-NH$_2$		5	2952	26.9	27.3	1.11	
20–30 nm		10	2869	27.5	28.3	1.09	
		20	2210	23.7	25.1	0.99	
		30	2005	22	23.8	0.99	
		40	1727	24	21.8	0.86	
UiO-66-Br		5	2890	20.1	18.1	1.49	
20–30 nm		10	2846	21.6	17.1	1.25	
		20	2416	19.3	16.3	1.53	
		30	2294	19	17.1	1.57	
		40	1441	23.6	20.8	1.03	
	6FDA-BisP	0	33.9		27.5		[91]
UiO-66		6	56.7		33.6		
50–100 nm		14	83.9		36.2		
		17	108		41.9		
		21	155		24.6		
	6FDA-ODA	0	25.9		20.6		
UiO-66		4	30.1		38		
50–100 nm		8	37.4		51.5		
		17	43.3		57		
		23	72		21.5		
	6FDA-DAM	0	997		29.2		
UiO-66		4	1283		29.6		
50–100 nm		8	1728		32		
		14	1912		30.9		
		21	2358		12.7		
	6FDA-DAM	0	1010 [c]		29.2		[92]
UiO-66		4	1290 [c]		29.6		
		8	1730 [c]		32.1		
		14	1915 [c]		31.2		
		21	2365 [c]		12.6		
UiO-66-NH$_2$		4	1295 [c]		29.2		
		8	1300 [c]		30.3		
		14	1345 [c]		29.9		
		21	1585 [c]		20.7		
UiO-66-NH-COCH$_3$		4	1081 [c]		30.3		
		8	1171 [c]		32.5		
		14	1266 [c]		33.1		
		21	1417 [c]		24.1		

[a] feed pressure = 9 bar; [b] feed pressure = 10 bar; [c] equimolar CO_2/CH_4 gas mixture.

In view of the reported data, UiO-66 appeared to be a promising inorganic phase to fabricate CO_2-selective hybrid membranes. Unlike the case of ZIF, the loading for UiO-66-based mixed matrix membranes has been limited to 40 wt.%, as agglomeration and poor polymer-fillers interface was observed at high loadings. In the case of unmodified particles, the CO_2 permeability was found to increase proportionally to the inorganic content for all of the investigated studies, but when considering the selective feature, an optimum is observed for a loading range between 10 and 20 wt.%. Amine modified UiO-66 (UiO-66-NH$_2$) showed typically better performance as compared to the pristine nanoparticles, which is mainly due to the enhanced CO_2-philicity. In general, surface modification led to improved polymer-particle interface, but for highly permeable polymers, this led to limited effect in terms of both selectivity and permeability.

2.3. Copper-Based MOFs

When compared to other metal organic frameworks, Cu-based MOFs offer an exceptional CO_2 uptake due to their high affinity with polar molecules. The presence of unsaturated open metal sites in Cu-based MOFs after activation is reported as an assisted mechanism in CO_2 sorption [93]. Comparison of the CO_2 adsorption capacity of two well-known MOFs containing the same ligand in their framework (Cu-BTC and Fe-BTC) showed that the Cu-BTC is characterized by a much larger CO_2 uptake (73.2 cm^3 g^{-1}, at room temperature and atmospheric pressure) when compared to the Fe-BTC (15.9 cm^3 g^{-1}) [94]. The results clearly pointed out the higher CO_2 affinity and interaction of Cu ions together with open metal sites, making Cu-based MOFs interesting for the fabrication of mixed matrix membranes for CO_2 separation.

Basu et al. [49] investigated the effect of $Cu_3(BTC)_2$ when embedded in Matrimid 9725 polymer phase. Upon the incorporation of the nanofiller, both CO_2 permeability (196%) and CO_2/CH_4 separation factor increased along with the inorganic content (Table 3). The overall increase of the separation performance was ascribed to the interactions between polymer and MOF and electrostatic interaction between the MOF and gas molecules, which leads to the existence of a competitive behavior. $Cu_3(BTC)_2$ was dispersed also in poly(2,6-dimethyl-1,4-phenylene oxide) (PPO), reducing the particles size via sonication from 50 to 6 μm [95]. Sonication was also reported to be able to improve the micropore volume of the nanoparticles and their dispersability within the polymeric matrix. By embedding 10 wt.% of $Cu_3(BTC)_2$ filler, the CO_2 permeability increased proportionally to the filler content, and the maximum enhancement was achieved for the smallest particles (26%, Table 3). The reduction of particles size showed a positive impact on the membrane selectivity: the improved compatibility with the polymer matrix prevented the selectivity drop observed as in the case of bigger particles. Abedini et al. [96] embedded $Cu_3(BTC)_2$ with a particle size of 100 nm in poly(4-methyl-1-pentyne) (PMP). By increasing the loading to 20 wt.%, they observed a simultaneous increase in CO_2 permeability (90%, Table 3) and selectivity (between 40 and 60% for the investigated gas pairs, Table 3). The observed variation was mainly attributed to a free volume increase. Interestingly, they also observed a reduction of the physical aging influence. Amine modification of $Cu_3(BTC)_2$ has also been reported as a possible approach to improve the CO_2 separation performance of a PEBAX 1657 [97]. In view of the H-bonding between the -NH_2 group and the polymeric chain, the modified nanoparticles showed better compatibility with the polymer phase. The CO_2 permeability increased proportionally to the loading (up to 100% increment, Table 3) similarly for both of the fillers, but better improvement of the CO_2/CH_4 selectivity was achieved upon amine-modification of the fillers. Interactions between the amine groups and the CO_2 have also been suggested to be responsible for the improved CO_2-philicity of the hybrid matrices.

Metal-organic polyhedral 18 (MOP-18) was also used to fabricate the hybrid membrane using Matrimid as polymer phase [98]. The inorganic content was increased up to 80 wt.%, but above 44 wt.% the samples' brittleness did not allow for the investigation of the transport properties via permeability testing. The CO_2 permeability increased along with MOP-18 content, even though a reduction in ideal selectivity was observed for both CO_2/N_2 and CO_2/CH_4. The permeability enhancement was attributed to increasing the number of alkyl chains, which improved the CO_2 solubility within the hybrid matrix. H_2 permeability was also measured and the addition of the nanoparticles increased the CO_2-philicity of the mixed matrix (Table 3).

Ahmadi et al. [6] synthesized a new class of Cu-based microporous metal-imidazolate framework (MMIF) and explored the separation performance of the mixed matrix membranes with 10 wt.% and 20 wt.% loading in Matrimid 5218 polymer matrix. The gas permeability showed a moderate increase (26%) along with the MOF content with limited effect on CO_2/CH_4 and CO_2/N_2 selectivity. The single gas permeation results revealed a flexible structure of MMIF, with the consequent formation of interfacial defects and voids. Interestingly, a significant enhancement of the separation factor was measured for mixed gas experiments (Table 3), which was mainly attributed to CO_2 competitive sorption within the hybrid matrix. Molecular simulation revealed that gas sorption was the dominant mechanism in the hybrid membrane, and the preferential CO_2 uptake into the MMIF pores limited the transport of other gases (CH_4 and N_2) through the MMIF's framework.

Zhang et al. [99] fabricated a Cu-based microporous metal-organic framework (Cu-BPY-HFS) and dispersed it in Matrimid 5218 up to 40 wt.% loading. The SEM images proved the good adhesion between MOF and Matrimid for loading up to 30%, but a higher amount of MOFs generated the formation of a poor particle-polymer interface. The CO_2 permeability increased along with the Cu-BPY-HFS content, and the variation was attributed to the 0.8 Å pore diameter and the presence of interfacial voids. Interestingly, the MOF was shown to have better affinity with CH_4 than CO_2, and enhanced CH_4 transport was observed in both pure and mixed gas tests. On the other hand, the CO_2/N_2 selectivity was negligibly affected by the presence of the MOF. As for the previous case, the addition of Cu-based MOF nanoparticles increased the CO_2-philicity, thus reducing the ability of the hybrid membranes to separate H_2 from CO_2.

Table 3. Gas separation performance of Cu-based MOFs used to prepared mixed matrix membranes (operating conditions ranging within 1–5 bar, 20–35 °C, unless differently specified).

Filler	Polymer	Loading (wt.%)	P_{CO_2} (Barrer)	α_{CO_2/N_2}	α_{CO_2/CH_4}	α_{CO_2/H_2}	Ref.
$Cu_3(BTC)_2$ 10 μm	Matrimid 9725	0	0.21 [a]		28.0		[49]
		10	0.3 [a]		30.0		
		20	0.41 [a]		31.0		
		30	0.64 [a]		32.5		
$Cu_3(BTC)_2$ 6 μm	PPO	0	68.9	16.1	16.2	0.92	[95]
		10	87.2	23.8	28.2	0.94	
$Cu_3(BTC)_2$ 100 nm	PMP	0	76.1	20.5	15.2	7.5	[96]
		5	88.3	22.2	17.1	8.1	
		10	103	23.7	19.2	9.2	
		15	124	25.4	22.7	10.7	
		20	144	28.6	24.3	12.2	
$Cu_3(BTC)_2$	PEBAX 1657	0	84.2		16.4		[97]
		5	91.4		17.7		
		10	102.7		19		
		15	128.8		20.5		
		20	167.3		19.5		
NH_2-$Cu_3(BTC)_2$		5	93		18.4		
		10	108.8		21		
		15	135.2		23.6		
		20	170.1		26.2		
MOP-18	Matrimid 5218	0	7.3	30.5	32.8	0.43	[98]
		23	9.4	27.6	23.2	0.53	
		33	14	22.9	21.8	0.63	
		44	15.6	26.0	16.4	0.70	
MMIF 50 nm 200 nm	Matrimid 5218	0	6.8	26.2	35.9		[6]
		10	8.1	27.3	36.9		
		20	8.6	27	34.6		
50 nm 200 nm	Matrimid 5218	0	8.0 [b]		38.3		
		10	9.7 [b]		81		
		20	10.1 [b]		88		
50 nm 200 nm	Matrimid 5218	0	7.1 [c]	32.3			
		10	8.2 [c]	38.9			
		20	11.7 [c]	58			
CU-BPY-HFS [d] 200–300 nm	Matrimid 5218	0	7.3	33.1	34.7	0.42	[99]
		10	7.81	32.5	31.9	0.46	
		20	9.88	31.9	27.6	0.59	
		30	10.36	33.4	27.4	0.51	
		40	15.06	30.7	25.5	0.56	

[a] Permeance (GPU); membrane thickness 40–65 μm; [b] Equimolar CO_2-CH_4 gas mixture; [c] Equimolar CO_2-N_2 gas mixture; [d] Cu-BPY-HFS: Cu–4,4′-bipyridine–hexafluorosilicate.

According to the results that were obtained from hybrid membranes containing different Cu-based MOFs, the high CO_2 uptake capacity leads to high CO_2 permeability. Although the increase in the gas transport is proportionally to the MOF content, a limited impact is observed on the selective feature of the hybrid membranes. Interestingly, this effect is more evident when the performance are evaluated through mixed gas permeation, which is presumably due to competitive sorption phenomena. In addition, amine modification appears to be a promising approach to improve the CO_2-philic nature of the mixed matrix membranes.

2.4. Materials Institute Lavoisier MOFs (MILs)

Material Institute Lavoisier (MILs) is a sub-family of MOFs that is based on trivalent metals strongly bonded to oxygen-anion-terminated linkers. MIL-53 (chemical formula: $M(OH)(O_2C-C_6H_4-CO_2$, $M = Al^{3+}$, Cr^{3+}) is made of dicarboxylate ligand interconnected by octahedral chains of $MO_4(OH)_2$ and has a 3D porous structure with one-dimensional (1D) diamond-shaped channels [100]. Furthermore, it is characterized by a pore limiting diameter of ~8.5 Å and surface area (Langmuir) of 1500 m^2/g [100,101]. Porous terephthalate MIL-53 showed a promising potential for H_2 storage and CO_2 capture [101–107]. CO_2 adsorption experiments showed that MIL-53 exhibited a two-step sorption isotherm, behavior that was associated to gate opening phenomena. Notably, even though this type of phenomena is typically observed in MOFs at low pressure, in the case of MIL-53, this happened for pressure above 5 bar, determining the two-step shape of the sorption isotherm [100]. Depending on the metal coordinate, different breathing mechanisms have been observed: upon dehydration, for example, MIL-53(Cr) and MIL-53(Fe) have open-pore and closed-pore structures, respectively.

Dorosti et al. [108] incorporated MIL-53 in Matrimid 5218 up to 20 wt.% loading. Strong interactions of CO_2 molecule with the oxygen atom of hydroxyl groups present in the MIL-53 structure and the breathing effect resulted in an enhancement of CO_2 permeability proportional to the MOF loading (Table 4). The CO_2/CH_4 selectivity showed a significant enhancement from 31 to 51.8 between the 10 and 15 wt.% loading. However, the formation of unselective voids at higher MIL-53 content led to a significant drop of the separation performance. In view of the breathing mechanism that is affecting the MOF framework, for pressure below 3 bar MIL-53 was found to be in its close-form, whereas at higher pressure an open-framework configuration was suggested. Higher MIL-53 contents (33.3 and 37.5 wt.%) in Matrimid were investigated by Hsieh et al. [109]. They investigated the effect of the reversible structure (closed or open form) on the transport properties. In this regard, MOF dehydrated with high temperature treatment (MIL-53-ht) and characterized by an open-pore structure was compared with as synthesized nanoparticles (MIL-53-as), which presented a closed-pore configuration. At a given loading (37.5 wt.%) MIL-53-ht showed a higher permeability when compared to the as synthesized MIL, but the selective features were significantly compromised in the open-pore configuration. CO_2/CH_4 selectivity as high as 90.1 for CO_2/CH_4 was reported for MIL-53-as (Table 4). The significant enhancement was due to the sieving effect produced by the partial blockage of the pores by the BDC linkers, which prevents the permeation of molecules with larger kinetic diameter (CH_4 and N_2). To further prove the effect of the pore structure, MMMs containing MIL-53-lt (activated at 50 °C) were shown as a framework transition from close pore form to open pore.

Abedini et al. [110] loaded amine-functionalized MIL-53 (Al) (100 nm size) in Poly(4-methyl-1-pentyne) (PMP) and prepared mixed matrix membranes with loading up to 30 wt.%. Addition of NH_2-MIL-53 into polymer matrix enhanced both CO_2 permeability and CO_2/CH_4 selectivity (Table 4), which is mainly due to improved CO_2 solubility in the hybrid matrix. At higher loading, the membrane performance overcame the Robeson upper bound for CO_2/CH_4 separation. In mixed gas, the same trend was observed for selectivity and permeability. However, lower separation performance (10% lower permeability and 30% lower selectivity) were observed when mixed gas conditions were investigated, which is possibly due to competitive sorption phenomena. Interestingly, it was observed that the addition of the porous nanoparticles increased the

CO_2-philicity of the hybrid membranes, decreasing the H_2/CO_2 selectivity that was observed for the pristine polymer. MIL-53 and amine functionalized NH_2-MIL-53 (Al) have also been dispersed in Poly(vinylidene fluoride) (PVDF) [111] and modified PVDF [112]. The modification of PVDF by means of KOH and $KMNO_4$ appeared to enhance the effect of the nanoparticle to a significant extent (Table 4). The CO_2 permeability doubled and the CO_2/CH_4 selectivity showed a notable enhancement (+50%) at 10 wt.% loading, with a minor effect being observed for modified and unmodified nanoparticles. In the case of pristine PVDF, a 50% enhancement of the CO_2 permeability was associated with small influence on the selective feature of the hybrid membranes.

Aiming at improving the interfacial interaction between MIL-53 and the polymer matrix, Tien-Binh et al. [113] introduced hydroxyl group into 6FDA-DAM polyimide backbone. 6FDA-(DAM)-(HAB) x:y copolymer (x and y denoted the copolymer ratio) containing hydroxyl groups facilitated the dispersion of MIL-53 (Al) and NH_2-MIL-53(Al). Single gas and mixed gas (CO_2/CH_4 50:50) separation performances were investigated for the mixed matrix membranes varying the copolymer ratio. Gas permeation characterization showed that the incorporation of pristine MIL-53 resulted in an increase in CO_2 permeability for both homopolymer and copolymers with increasing the MIL loading, with the effect becoming more influential for the low permeable samples (i.e., increasing the DAM/HAB ratio, Table 4). The formation of interfacial voids is suggested to be responsible for the observed variations. On the other hand, when modified MIL-53 was used a different behavior is observed: a minimum was observed for the CO_2 permeability at 15 wt.% loading, whereas the CO_2/CH_4 selectivity was optimized at 10 wt.% loading. In view of the favorable interactions between the hydroxyl and the amine group, the increase in selectivity became more significant at a higher DAM/HAB ratio, also reducing the negative effect on the CO_2 transport across the hybrid membranes. SEM images supported this observation. Zhu et al. [114] investigated the performance of thin film hollow fiber mixed matrix membranes filled with post-modified MIL-53 (P-MIL-53). Asymmetric hollow fibers (Ultem) coated were used as support and PDMS containing different MIL-53 content was used as selective layer. The obtained results showed that the membranes containing 15% P-MIL-53 showed the best performance: the CO_2 permeance was improved from 30 GPU to 40 GPU when compared to hollow fiber membranes coated with only pure PDMS. At 15% loading, the ideal selectivity increased from 23.3 to 28.1 for CO_2/N_2 and from 27 to 32 for CO_2/CH_4. This was mainly attributed to the strong affinity with CO_2 due to dipole-quadrupole interaction of CO_2 molecules with NH_2 groups in the MOF. At 20 wt.% loading, a decrease in CO_2/N_2 and CO_2/CH_4 ideal selectivity was observed, which is mainly ascribed to particle agglomeration.

MIL-101 is another MOF from the MILs' family, widely studied for gas separation application [115]. The MIL-101 framework is composed of chromium atoms making an octahedral framework with oxygen atoms and 1.4-benzene dicarboxilate (BDC) ligands. The rigid terephthalate ligand together with trimeric chromium octahedral clusters provides window aperture of 8.5 Å and accessible large cages. The gas sorption analysis showed that a Langmuir surface area of 5900 m^2/g [115]. Similar to MIL-53, the removal of water molecules from the structure leaves unsaturated open metal sites in the MIL-101 structure [101].

Naseri et al. [116] recently reported the gas separation performance of hybrid Matrimid membranes containing MIL-101 (Cr) up to 30 wt.% loading (10 bar and 35 °C). The presence of MIL-101 in the polymer matrix enhanced the CO_2 permeability (Table 4), with the main contribution coming from the increase in CO_2 solubility within the hybrid matrices. The ideal CO_2/CH_4 and CO_2/N_2 selectivity showed a maximum at low loading (10 wt.%) and the drop of selective features at higher loading is mainly attributed to the presence of non-selective voids at the polymer/particles interface. The effect of addition of MIL-101(Cr) on the separation performance of a blend of Matrimid and PVDF was investigated by Rajati et al. [117]. 3 wt.% PVDF in Matrimid was selected as the most suitable blend composition for CO_2/CH_4 separation, which showed higher CO_2 permeability (28%) and selectivity (22%) when compared to pristine Matrimid. The embedment of 10 wt.% MIL-101 showed a similar effect on both the pristine polymer matrix and the polymer blend, with about 60%

increase in CO_2 permeability and 40% higher selective features. The simultaneous enhancement of permeability and selectivity suggested the presence of a proper interface morphology. Additionally, the electrostatic interaction of functional groups in MIL-101 with CO_2 resulted in better affinity and higher solubility.

MOFs-derived porous carbons (PC) based on MIL-101(Cr) and MIL-53(Al) were prepared by soaking the MOFs into NH_4OH and carbamide, followed by calcination at 800 °C [118]. The carbonized MOFs were embedded into PPO-PEG at a loading range between 5 and 25 wt.%. For both nanoparticles, limited changes in CO_2 permeability were observed up to 20 wt.% loading, but a marked improvement was observed at 25 wt.% loading, achieving promising permeability values (Table 4). MIL-101(Cr)-PC showed a better performance (1896 Barrer) when compared to MIL-53(Al)-PC (1266 Barrer). The selectivity showed also an improvement, and the optimum at 20 wt.% MIL content clearly suggested that higher loading probably generated interfacial voids and particle agglomeration. However, unlike the effect on permeability, MIL-53(Al)-PC showed higher selective feature when compared to MIL-101(Cr)-PC.

An interesting approach to optimize the performance of mixed matrix membranes is represented by the use of mixed MOFs [119]. A mixture of MIL-101/ZIF-8 was homogenously dispersed in PSF and no agglomeration was observed. The MMMs performance showed an enhancement in CO_2 permeability as a function of filler loading, and the simultaneous presence of both MIL and ZIF nanoparticles showed a synergetic effect. At 35 wt.% MOF loading, the CO_2 permeability was significantly increased (six-fold) when compared to the pristine PSF, from 5 Barrer to 30 Barrer. This was explained as increasing free volume of polymer associated to a disruption of the polymeric chains, together with the larger pore size of MIL-101. At an intermediate loading, 16 wt.%, the CO_2/CH_4 separation factor was increased from 23 to 40 as compared to pristine PSF. Higher loading, 35 wt.%, led to a selectivity drop, due to poor interface morphology. The authors suggested that the coexistence of ZIF-8 and MIL-101 improved the dispersion and avoided agglomeration at low particles loading.

Finally, an interesting use of MILs as MOF scaffold (MS) has been proposed by Xie et al. [120], where the separation performance of membranes obtained from MOFs and PEG (MSP) were investigated for post combustion CO_2 capture (CO_2/N_2 10/90). Firstly, the MS membranes were fabricated on a support; then, coatings with different PEG concentration were applied to prepare the MSP membranes. The MS membranes showed extremely high CO_2 permeance (85000 GPU), but no selective feature. Upon the application of PEG coating (PEG concentration > 0.6 mmol/5mL aqueous solution) suitable selectivity value (>30) were achieved, maintaining high CO_2 permeability (>2700 Barrer). It was suggested that the coated polymer provides a defect free membrane and a shorter path for CO_2 transport.

Table 4. Gas separation performance of MIL-based mixed matrix membranes (operating conditions ranging within 1–5 bar, 20–35 °C, unless differently specified).

Filler	Polymer	Loading (wt.%)	P_{CO2} (Barrer)	α CO2/N2	α CO2/CH4	α CO2/H2	Ref.
	Matrimid 5218	0	6.2		28.2		[108]
MIL-53 (Al)		5	6.8		29.6		
123–466 nm		10	7.45		31		
		15	12.43		51.8		
		20	14.52		15.1		
	Matrimid 5218	0	8.4	33.6	39.4	0.33	[109]
MIL-53-as [a]		37.5	40	95.2	90.1	0.55	
MIL-53-ht [a]		33.3	26.6	42.9	45.7	0.50	
50–100 nm		37.5	51	28.3	47.0	0.60	

Table 4. *Cont.*

Filler	Polymer	Loading (wt.%)	P_{CO_2} (Barrer)	α_{CO_2/N_2}	α_{CO_2/CH_4}	α_{CO_2/H_2}	Ref.
NH$_2$-MIL-53 (Al) 110 nm	PMP	0	98.74		8.72		[110]
		5	107.32		11.85		
		10	118.74		12.59		
		15	139.56		15.72		
		20	164.78		18.46		
		25	203.44		20.18		
		30	226.37		20.36		
MIL-53 100 nm	PVDF	0	0.92	16.3	21.3		[111]
		5	1.21	16.3	21.2		
		10	1.55	16.2	21.0		
NH$_2$-MIL-53(Al) 100 nm		5	1.11	17.3	23.1		
		10	1.41	19.5	26.0		
MIL-53 100 nm	m-PVDF [b]	0	1.2		27.9		[112]
		5	1.75		35.8		
		10	2.45		39.6		
NH$_2$-MIL-53 100 nm		5	1.69		37.6		
		10	2.24		43.2		
MIL-53 (Al) 190–340 nm	6FDA–(DAM)	0	316.6 [c]		9.76		[113]
		10	331.9 [c]		10.19		
		15	354.0 [c]		11.46		
	6FDA–(DAM)–(HAB) 2:1	0	115.7 [c]		21.65		
		10	124.2 [c]		24.62		
		15	134.5 [c]		26.96		
	6FDA–(DAM)–(HAB) 1:1	0	46.8 [c]		34.39		
		10	55.3 [c]		37.15		
		15	63.0 [c]		40.76		
	6FDA–(DAM)–(HAB) 1:2	0	19.6 [c]		43.1		
		10	33.2 [c]		47.13		
		15	42.6 [c]		48.83		
NH$_2$-MIL-53 (Al) 100–200 nm	6FDA–(DAM)	0	316.2 [c]		9.77		
		10	308.9		13.63		
		15	290.7 [c]		14.77		
		20	299.8 [c]		8.86		
	6FDA–(DAM)–(HAB) 2:1	0	115.7 [c]		21.81		
		10	112.1 [c]		43.63		
		15	105.7 [c]		36.13		
		20	122.1 [c]		29.31		
	6FDA–(DAM)–(HAB) 1:1	0	47.4 [c]		34.54		
		10	43.7 [c]		77.72		
		15	44.6 [c]		64.54		
		20	54.7 [c]		35.68		
	6FDA–(DAM)–(HAB) 1:2	0	24.6 [c]		53.86		
		10	20.0 [c]		86.81		
		15	21.9 [c]		96.36		
		20	31.9 [c]		55.9		
P-MIL-53 500 nm	PDMS	0	30	23.3	27.0	0.22	[114]
		5	33.3	24.5	28.8		
		10	36.0	25.8	30.5	0.24	
		15	40.3	28.1	32.1		
		20	42.3	27.5	28.4		

Table 4. *Cont.*

Filler	Polymer	Loading (wt.%)	P$_{CO2}$ (Barrer)	α $_{CO2/N2}$	α $_{CO2/CH4}$	α $_{CO2/H2}$	Ref.
	Matrimid 5218 [d]	0	4.44	34	35		[116]
MIL-101(Cr)		10	6.95	52	56		
~1000 nm		15	5.7	44	47		
		20	5.85	42	37		
		30	7.99	47	44		
	Matrimid 5218 [f]	0	7.33		34.9		[117]
MIL-101(Cr)		10	12.01		52.21		
	Matrimid/PVDF [f]	0	9.42		42.81		
MIL-101(Cr)		10	14.87		62		
	PPO-PEG [c,e]	0	657		18.42		[118]
MIL-53(Al)-PC		5	684		25.51		
200–250 nm		10	723.6		29.23		
		15	763		35.78		
		20	789		40.39		
		25	1266		31.53		
MIL-101(Cr)-PC	PPO-PEG [c,e]	0	657		19.26		
50–100 nm		5	771		22.93		
		10	874		26.61		
		15	952		30.46		
		20	1056		34.66		
		25	1896		29.24		
	PSF	0	5		23		[119]
MIL-101		8	8		21		
110–400 nm		16	8.9		24		
		24	18.1		28		
ZIF-8		0	5		23		
75–100 nm		8	10		35		
		16	14		22		
		24	24		24		
MIL-101/ZIF-8		0	4.7		23		
		8	10.6		36		
		16	14.2		40		
		24	24		26		
		35	29.6		24		

[a] as = as synthesized, "ht" = high temperature treated (300 °C); [b] m-PVDF = modified poly(vinylidene fluoride); [c] mixed gas conditions; [d] feed side pressure = 10 bar; [e] PPO-PEG = polyphenylene oxide-polyethylene glycol; [f] feed pressure = 7 bar; [g] Permeance (GPU); [h] MSxPy: MOFs Scaffold.

Similar to the previous MOFs, the MILs' family also represents a group of nanoporous particles that is suitable for the development of mixed matrix membranes for CO_2 capture. The CO_2 permeability is frequently found to increase along with the loading, but a loading range between 10 and 15 wt.% appears to be the one that is able to optimize the selective feature of the hybrid membranes. Favorable interactions with the polymeric matrix act in the direction of enhancing the CO_2-philicity of the mixed matrix membranes. The closed-pore structure appears to be the most suitable one for the achievement of improved separation performance; whereas, the open-pore structure is expected to enhance the gas transport through the hybrid matrix, thus possibly compromising the selectivity.

2.5. Other MOFs

Fe-BTC is reported to be a low cost and water stable MOF type that exhibits a pore size between 5.5 and 8.6 Å and a relatively higher surface area when compared to its Cu counterpart. Despite the lower uptake capacity when compared to Cu-BTC, the presence of a large number of coordinatively unsaturated sites and high water stability make the MOF a suitable candidate for the fabrication of mixed matrix membranes for CO_2 separation. Fine Fe-BTC particles were dispersed in Matrimid 5218 matrix to prepare hybrid membranes, and the effect of the fillers on the gas transport properties and plasticization behaviour were investigated [121]. While limited effects were observed in single gas tests (Table 5), under mixed gas and high pressure (~40 bar) conditions, the CO_2 permeability increased by 30% and CO_2/CH_4 selectivity by 62% when compared to the neat polymer. The chain rigidity of the MOF also contributed to enhance the plasticization resistance of the hybrid membrane up to 20 bar. The effect of Fe-BTC filler on the transport properties of Matrimid has also been studied by Rita et al. [122]. The study revealed that gas diffusivity changes with increasing temperature dominated the drop in solubility, leading to an overall increase in CO_2 permeability from 94.2 Barrer at 303 K to 217.9 Barrer at 353 K with a 30 wt.% MOF loading. Interestingly, the CO_2/N_2 selectivity of the matrix increased on a similar scale with temperature increase. The effect of Fe-BTC on rubbery PEBAX 1657 for gas permeation was studied by Dorosti and Alizadehdakhel [123]. Both single gas and mixed gas (CO_2/CH_4) tests revealed a four-time increase in CO_2 permeability when compared to the neat polymer (Table 5). The gas selective feature of the hybrids showed a minor increase as compared to the pristine polymer, but a significant drop is observed at 40 wt.% loading due to the formation of non-selective voids. Differently from what has been observed for the glassy polyimide, the increase in feed pressure led to plasticization phenomena, and consequently to a drop in selectivity.

A new sorption selective, chemically stable, fluorinated MOF NbOFFIVE-1-Ni (KAUST-7) was developed by Cadiau et al. [124]. KAUST-7 showed an apparent pore size of 4.75 Å and a CO_2 sorption capacity of 2.2 mmol/g at 25 °C and 1 bar. Recently, Chen et al. [125] synthesized nanosized KAUST-7 crystals by novel co-solvent synthesis method (Figure 7) and dispersed them in 6FDA-Durene matrix. The CO_2 permeability increased along with the loading from 750 (pristine polymer) to 1038 (33 wt.% loading) Barrer (Table 5). The selectivity marginally increased due to both increase in solubility selectivity and diffusivity selectivity. Additionally, interactions between the organic ligand and the groups of 6FDA increased compatibility, leading to enhanced plasticization resistance up to 10 bar, with a minor reduction in CO_2/CH_4 selectivity of 33% MOF loaded matrix.

Bimetallic MOFs, like $Mg_2(dobdc)$, contain many open metal sites along the pore walls facilitating a selective adsorption and transport of CO_2. Bae and Long [126] developed a facile synthesis method to produce 100 nm primary crystals of $Mg_2(dobdc)$ and successfully incorporated them in three different polymer matrices: PDMS, crosslinked-PEO, and 6FDA-TMPDA (polyimide). The study revealed that the MOF had a negative effect on the gas transport through the rubbery polymers (Table 5), possibly due to the plugging of the MOF pores by the rubbery polymer chains. On the other hand, a simultaneous enhancement of both CO_2 permeability and CO_2/N_2 selectivity was observed for the glassy polyimide (Table 5). It was shown that the variation was mainly associated to the increase in CO_2 solubility, with minor effects on the gas diffusion through the selective layer. A similar study by Smith et al. [127] proved that the addition of $Mg_2(dobdc)$ to 6FDA-Durene increased the permeability of CO_2, N_2, H_2, and CH_4 due to the increase in diffusivity of the penetrants. It was observed that the MOF particles further increased the brittleness of the films due to densification. By changing the coordination site, $Ni_2(dobdc)$ was fabricated and it was found to improve the mechanical robustness, owing to smaller primary particle size. Both bimetallic MOFs were found to improve the performance in separations governed by diffusivity differentials, like H_2/CH_4 and H_2/N_2 when compared to CO_2/CH_4 and CO_2/N_2 separations that require both solubility and diffusivity enhancement.

Table 5. Gas separation performance of different MOFs (Fe(BTC), KAUST-7, $Mg_2(dobdc)$) used to prepared mixed matrix membranes (operating conditions ranging within 1–5 bar, 20–35 °C, unless differently specified).

Filler	Polymer	Loading (wt.%)	P_{CO2} (Barrer)	$\alpha_{CO2/N2}$	$\alpha_{CO2/CH4}$	$\alpha_{CO2/H2}$	Ref.
	Matrimid 5218	0	9		25		[121]
Fe(BTC)		10	9.5		27.5		
		20	10.8		28		
		30	13.1		29.5		

Table 5. *Cont.*

Filler	Polymer	Loading (wt.%)	P_CO2 (Barrer)	α CO2/N2	α CO2/CH4	α CO2/H2	Ref.
Fe(BTC) 10–20 um	Matrimid 5218 [a]	0	14.6	4.4			[122]
		10	84.9	43.5			
		20	91.2	15.4			
		30	217.9	23.1			
Fe(BTC)	Pebax 1657	0	70.67		18.4		[123]
		5	80.79		19.3		
		10	82.32		19.4		
		15	89.63		20.8		
		20	98.32		22.2		
		25	148.44		21.9		
		30	402.69		21.5		
		40	425.5		12.3		
		0	60.35 [b]		16.9		
		10	70.11 [b]		17.6		
		20	85.28 [b]		19.3		
		30	329.7 [b]		20.5		
		40	345.4 [b]		13.1		
KAUST-7 80 nm	6FDA Durene	0	759.7 [b]		34.7		[125]
		11	895.7 [b]		36.2		
		22	966.9 [b]		37.0		
		33	1038.1 [b]		37.6		
Mg_2(dobdc) 100 nm	PDMS	0	3100.0	9.5			[126]
		20	2100.0	12			
	XLPEO	0	380.0	22			
		10	250.0	25			
	6FDA-TMPDA	0	650.0	14			
		10	850.0	23			

[a] Temperature = 80 °C; [b] Mixed gas CO_2/CH_4 10/90.

Figure 7. Fine-tuning crystal size of KAUST-7 by varying ethanol-water (solvent) ratios in synthesis solution: (**a**) pure water, ethanol/water ratio of (**b**) 0.46, (**c**) 0.82, (**d**) 1.2, (**e**) 1.5, (**f**) 2.2, (**g**) 2.7, (**h**) 6.3, and (**i**) pure ethanol [125], with copyright permission from © 2018 Elsevier.

3. Porous Organic Frameworks (POFs)

Metal-organic frameworks have drawn considerable attention for their tunable chemistry and gas separation and storage performance, many MOFs suffer from the lack of chemical and physical stability. In addition, their limited sorption capacity and the presence of heavy metal ions in their framework have posed obstacles for their prospective applications [128]. Recently, a new class of porous materials known as porous organic frameworks (POFs) has attracted great attention as an alternative to MOFs. POFs can be either crystalline, such as covalent organic frameworks (COFs), or amorphous with uniform pore diameter, such as porous aromatic frameworks (PAFs) [129,130]. Due to the entirely organic structure, POFs ensure good adhesion with organic polymer phase and display better chemical compatibility [129]. PAF-1 (Figure 8) was synthesized and characterized for the first time in 2009, with the scope of exploring its potential as adsorbent [131]. PAFs have a local diamond-shape with tetrahedral bonding of tetraphenylene methane in their main building block. The exceptional surface area (Langmuir surface area of 7100 m^2/g) of PAFs has shown excellent sorption capacity for hydrogen and carbon dioxide (i.e. 1300 mg/g CO_2 uptake at 25 °C and 40 bar). Furthermore, they are characterized by super hydrophobicity, enhanced adsorption enthalpies, and delocalized charged surface [128,131]. Thermal analysis of PAFs exhibited that the structural integrity remained intact up to 520 °C in air and after water boiling point for seven days [131]. The pore size distribution of PAF-1 displays a pore diameter of 1.4 nm, which can be tuned via activated carbonization to 0.79, 0.93, 0.64, and 0.6 nm while using KOH, NaOH, CO_2, and N_2 as an activation agent, respectively [132]. Furthermore, a Monte Carlo simulation study suggested that a nitrogen-doped PAF (NPAF-11) containing imidazolic group improves the CO_2 uptake more than 130% when compared to PAF-1 [133].

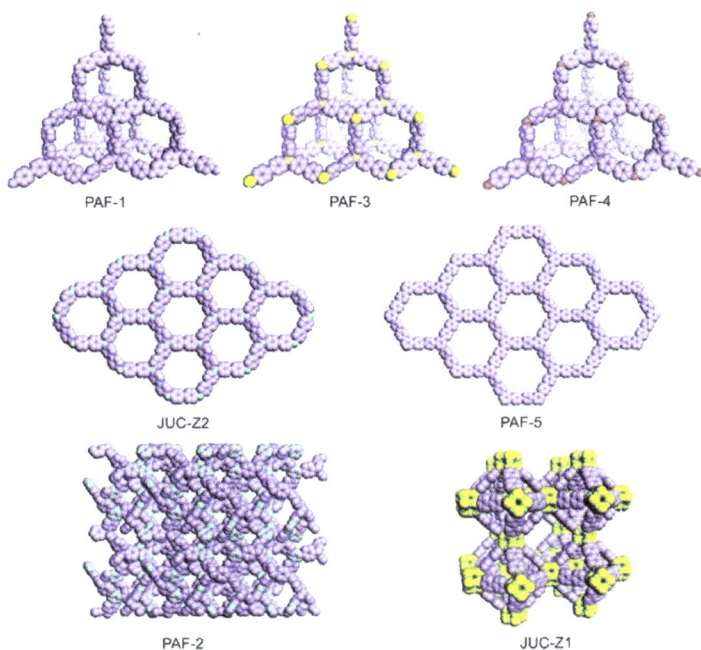

Figure 8. Structure model of synthesized and simulated porous aromatic frameworks. Atom colors: C = purple, N = blue, Si = yellow, O = green, Ge = brown [134], with copyright permission from © 2012, Royal Society of Chemistry.

The non-equilibrium nature of glassy polymers makes them subject to physical aging, which tends to reduce their fractional free volume over time, and thus, the gas permeability coefficient. Porous organic frameworks have been reported to have the ability to play as an anti-aging filler, as they can freeze the nanostructural morphology, slowing down the aging process to a significant extent. Lau et al. [135] embedded PAF-1 (10 wt.% loading) into three high free volume glassy polymers including PTMSP, PIM-1, and PMP in order to explore the influence of porous organic fillers on aging process of these polymers. Over a period of eight months (240 days), the CO_2 permeability of pristine PTMSP dropped from 20,000 Barrer to a value of 12,500 Barrer (37% drop). The hybrid membrane containing 10 wt.% PAF-1 showed a higher CO_2 permeability (approximately 25,000 Barrer), which dropped of only 7% over the investigated period. Similar effects were observed for PIM-1 and PMP (Table 6). Interestingly, the CO_2/N_2 selectivity improved with aging, similarly to the pristine matrix. Following a similar goal, Volkov et al. [136] embedded PAF-11 in PTMSP membrane (1–10 wt.% loading) and monitored the variation of the transport properties over time through single gas permeation experiments. Initially, the addition of PAF-11 nanoparticles corresponded to an increase of the gas permeability of PTMSP, with a negligible effect on the selective features of the membranes. Long-term exposure to high temperature showed that the presence of the PAF nanoparticles helped in improving the mechanical stability of the PTMSP matrix: pristine PTMSP could not withstand more than 200 h exposure at high temperature, whereas the hybrid matrixes were tested up to 510 h, showing good mechanical properties. Furthermore, the membrane with 10% PAF-11 loading showed a limited drop of the CO_2 permeability (30%), with stable performance over a period of more than 300 h.

Functionalization of PAF-1 has been reported as an effective method to improve the CO_2 permeability in hybrid membranes [137]. The presence of functional groups (NH_2, SO_3H, C_{60} nanoparticles, and Li_6C_{60} composites) added to PAF-1 affected the CO_2 sorption capacity, mainly due to the affinity of polar functional groups with CO_2. Particularly promising is the introduction of Li_6C_{60} functionality, which is able to provide additional CO_2 sorption sites that are associated to the lithium, also increasing the PAF-1 surface area (from 3760 to 7360 m^2 g^{-1}). The CO_2 permeability of PTMSP increased from 30,000 to 55,000 Barrer, and the effect of aging was limited to a 10% decrease over a period of 365 days for a 10 wt.% of PAF-1-Li_6C_{60} loaded in PTMSP. CO_2/N_2, and CO_2/CH_4 selectivity were affected by the addition of the nanoparticles and by the physical aging to a limited extent (Table 6). Mitra et al. [138] studied the influence of a hypercrosslinked (HPC) nanofillers on the performance of PIM-1. PIM-1 membrane, prepared using dichloromethane as solvent, showed a CO_2 permeability of 2258 Barrer, which dropped to a value of 1109 Barrer after 150 days. A similar trend was also observed when chloroform was used as solvent (Table 6). The addition of HCP into PIM-1 reduced the effect of physical aging for the samples prepared with different solvents, but at high loadings, the selectivity was negatively affected by the presence of the nanoparticles. Interestingly, the addition of HCP was found to prevent membrane swelling in the presence of ethanol.

Table 6. Gas separation performance of porous organic frameworks (POFs)-based mixed matrix membranes (operating conditions ranging within 1–5 bar, 20–35 °C, unless differently specified).

Filler	Polymer	Loading (wt.%)	P$_{CO2}$ (Barrer)	α $_{CO2/N2}$	α $_{CO2/CH4}$	α $_{CO2/H2}$	Ref.
	PTMSP – 0 d	0	20,000	8.7			[135]
PAF-1		10	25,000	8.1			
	PTMSP – 240 d	0	12,400	9.8			
PAF-1		10	23,200	9.6			
	PIM-1 – 0 d	0	4000	15			
PAF-1		10	15,000	12			
	PIM-1 – 240 d	0	1700	19			
PAF-1		10	15,000	19			
	PMP– 0 d	0	6500	10.5			
PAF-1		10	11,500	9.4			
	PMP – 240 d	0	3500	11			
PAF-1		10	10,500	9.4			

Table 6. *Cont.*

Filler	Polymer	Loading (wt.%)	P_{CO_2} (Barrer)	α_{CO_2/N_2}	α_{CO_2/CH_4}	α_{CO_2/H_2}	Ref.
	PTMSP	0	30,000	5.6			[136]
PAF-11		1	38,000	5.9			
		5	37,000	5.8			
		10	34,000	5.6			
	510 hours	1	20,000	7			
		5	19,500	6.8			
		10	23,500	6.3			
	PTMSP	0	30,000	5.9	2.3		[137]
PAF-1		10	35,500	5.7	2.3		
PAF-1-NH$_2$		10	43,000	5.9	2.2		
PAF-1-SO$_3$H		10	32,500	5.7	2.3		
PAF-1-C$_{60}$		10	33,000	5	2.1		
PAF-1-Li$_6$C$_{60}$		10	55,000	5.4	2		
	Aged	0	8000	8.8	5.3		
PAF-1		10	28,000	7.4	3.1		
PAF-1-NH$_2$		10	29,000	7.5	3.6		
PAF-1-SO$_3$H		10	23,500	6	2.6		
PAF-1-C$_{60}$		10	15,000	8.3	5		
PAF-1-Li$_6$C$_{60}$		10	50,600	9	3.9		
	PIM-1 - CH$_2$Cl$_2$	0	2258	24			[138]
HCP		5.7	4690	17.6			
		16.67	5103	13.1			
		21.3	6331	14.1			
	150 d	0	1109	4.2			
HCP		5.7	3616	19.7			
		21.3	5060	16			
	PIM-1 - CHCl$_3$	0	2660	22.3			
HCP		4.6	4313	19.8			
		9.1	4700	19.3			
		16.67	10,040	17.1			
	150 d	0	1225	21.5			
HCP		4.6	1857	22.4			
		9.1	2043	22.2			
		16.67	4165	21.8			

Despite the limited amount of investigations, POFs appeared to be promising materials for the fabrication of CO_2 separation applications. The main advantage they add to polymeric materials is the significantly reduced physical aging phenomena; therefore, they are of interest for high free volume polymers. However, even though they are characterized by high CO_2 uptake, their addition can increase the CO_2 permeability (even in high free volume polymers), but has a limited or negligible effect on the selective feature of the hybrids. Interestingly, the young modulus has been reported to benefit from the addition of PAFs [136].

4. Zeolites

Zeolite molecular sieves are a class of aluminosilicate crystals that have been studied extensively and are one of the most widely reported porous materials for CO_2 capture because of their physiochemical properties [139,140]. The pore size of zeolites varies from 4 Å to 1.2 nm and their frameworks are formed by interconnecting channels. The molecular sieving nature coupled with the strong dipole-quadrupole interaction with carbon dioxide make zeolites promising candidates for carbon capture. Si and Al derived from silicate compounds are the main building block of zeolites nanoparticles. The morphology is controllable by varying the Si and Al content, as changes in the Si/Al ratio lead to the electrostatic charge variation, resulting in different pore sizes distribution and adsorption capacities [139]. The thermal and chemical stability of zeolites can be improved by increasing the Si content [139], even though the zeolites do not provide the level of tenability offered by MOFs [129,140]. Zeolites of interest for CO_2 capture applications are classified into three main categories: zeolites with small pore size (Linde Type A, LTA), medium pore size (Mordenite Framework Inverted, MFI), such as ZSM-5, and large pore size (Faujasite, FAU). Extensive studies have been dedicated to ensure the good adhesion between zeolites and polymer phases, as interfacial defects and voids between the organic and inorganic phases frequently resulted in the poor separation performance of the hybrids [141]. Unlike MOFs, zeolites structure is rigid and the pore dimensions are generally fixed. However, their activation by calcination may have detrimental impact on their framework integrity. The absence of accessible open metal sites (hidden by oxygen ions in the zeolite structure) is responsible for a lower CO_2 uptake when compared to MOFs [141]. The mechanism typically used to describe the transport of light penetrants through zeolites is solution-diffusion [142]. Extensive studies have been dedicated to the incorporation of zeolites in hybrid membranes for CO_2 separation [142]. Nevertheless, the research is still extremely active, and many studies on hybrid membranes for CO_2 applications employing zeolites have been reported in recent years.

Hoseinzadeh Beiragh et al. [143] investigated the effect of ZSM-5 loading on the CO_2/CH_4 separation performance of PEBAX-based membranes. The single gas permeation results revealed that an optimum for CO_2 permeability is achieved at low zeolite content (5 wt.%, Table 7), whereas the CO_2/CH_4 selectivity increased proportionally to the zeolite loading, achieving a 67% enhancement when compared to the pristine polymer. The sieving effect of zeolite (pore diameter 5.4 Å) has been suggested to be the main reason for the enhancement of the separation performance, and the decrease in fractional free volume was identified as explanation of the permeability drop at high zeolite contents. Interestingly, at higher feed pressure (up to 5 bar) the beneficial influence of zeolites on the mixed matrix performance is significantly reduced. Contrasting results were obtained when ZSM-5 have been embedded in a glassy polyimide (Matrimid 5218) [144]. In this case, the permeance increased along with the particle loadings (from 5 to 21 GPU), whereas the selectivity showed a 75% decrease. The results have been mainly associated to poor compatibility between ZSM-5 and Matrimid, which resulted in particles agglomeration and the presence of interfacial voids already at low loadings.

Zeolite 13X have been used by Bryan et al. [145] to prepare hybrid membranes based on PEBAX 1657. When compared to ZSM-5, X zeolites are characterized by a larger pore size, between 11 and 14 Å [146]. The addition of 15 wt.% nanoparticles in the polymer matrix led to the improvement of both CO_2 permeability (from 81 to 114 Barrer) and CO_2/N_2 selectivity (from 41 to 47), suggesting the achievement of proper interface morphology between the particles and the polymer matrix. Surya Murali et al. [147] have also prepared mixed matrix membrane using PEBAX 1675 as polymer phase. Zeolite 4A was embedded up to 30 wt.% in the polymeric matrix, showing agglomeration at higher loading. A 3-fold enhancement of the CO_2 permeability was observed with increase in inorganic loading, but the selective feature showed an optimum between 5 and 10 wt.%, which is possibly due to the interfacial voids formation. Zhao et al. [148] fabricated mixed matrix embedding up to 50 wt.% SAPO-34 zeolite (pore diameter 3.8 Å) in PEBAX 1657. The CO_2 permeability increased proportionally to the inorganic content, achieving three-fold enhancement when compared to the pristine polymer. The consistency with the Maxwell model prediction also suggested the achievement of a proper polymer-filler interface. However, the CO_2-selective feature of the hybrids were negligibly affected (Table 7). Interestingly, even though SAPO-34 inorganic membranes own impressive CO_2/H_2 selectivity [149], the performance of the pristine PEBAX 1675 were negligibly affected for the entire loading range investigated. Rezakazemi et al. [150] investigated the influence of 4A zeolite on the transport properties of polydimethylsiloxane (PDMS). The hybrid membranes showed a proper polymer-fillers interface up to 50 wt.% loading. Interestingly, a significant H_2-sieving effect was observed for the fabricated membranes: H_2 permeability increased along with the inorganic content, whereas both CO_2 and CH_4 transport was hindered. The pristine PDMS was found to be CO_2-selective for H_2 separation, but at 20 wt.% 4A loading, the hybrid material showed H_2—selective feature, suggesting that the membrane shifted from being solubility-driven to a condition where the diffusion coefficient dominates the gas transport.

Membranes **2018**, *8*, 50

Recently, Atalay-Oral et al. [151] proposed a comparative study about the effect of different zeolites on the transport properties of polyvinylacetate (PVAc). They compared four different zeolites: 4A (pore size 4.2 Å), Ferrierite (pore size 4.2 Å), 5A (pore size 5.2 Å), and Silicalite-1 (pore size 5.5 Å). For all of the different fillers, the CO_2-selective features of the mixed matrix membranes were increased. Ferrierite showed the better improvement in terms of performance: the selective feature (both CO_2/CH_4 and CO_2/N_2) increased proportionally to the inorganic content (Table 7), whereas the permeability increased at 20 wt.% loading, but negligible differences were observed at higher inorganic content. The authors suggest the strong interactions between Ferrierite cations and CO_2 molecules to be the main reason the superior performance of the Ferrierite-based hybrid membranes. Another study compared the performance of Zeolite A (Si/Al = 1) and zeolite ITQ-29 (Si/Al = ∞) when embedded in PTMSP [152]. Surprisingly, when single gas tests were performed on a hybrid membrane containing 20 wt.% Zeolite A loading, a 35% drop in CO_2 permeability was observed being combined with a 70-fold enhancement of the CO_2/N_2 selectivity (Table 7), surpassing the Robeson's upper bound. The extraordinary performance was attributed to the molecular sieving ability of the nanoparticles and to the achievement of a proper interface morphology. A much lower improvement was observed in the case of ITQ-29 zeolite, which is mainly due to the poor polymer-zeolite interactions and consequently interfacial voids formation. These results highlighted that choosing the proper Si/Al ratio is extremely important in the design of hybrid membranes, as it directly affects the organic/inorganic interfacial morphology. Nevertheless, the same authors reported that under mixed gas conditions, the separation factor of the Zeolite A/PTMSP membranes appeared to be lower (5) when compared to the ideal selectivity (63) [153]. The authors concluded that the influence of the diffusivity selectivity dominates the transport, rather than the preferential sorption capacity in the mixed matrix.

As previously reported for MOFs and POFs, surface functionalization of zeolites is reported as a successful approach to improve the polymer-particles compatibility, and, thus, the membrane performance. The presence of unselective interfacial voids at the interface between zeolite 4A and PSF determined a significant drop of the CO_2/CH_4 separation efficiency (Table 7), without enhancing the CO_2 permeability to a significant extent [154]. However, the functionalization of the zeolites particles with $MgCl_2$ and NH_4OH resulted in increased selectivity up to 30 wt.% loading, with a limited effect on the CO_2 permeability. Similarly, zeolite 5A have also been modified with Mg-based moieties to improve the adhesion with the polymer chain in Matrimid-based membranes [155]. Surface treatment of zeolite with $Mg(OH)_2$ improved both CO_2 permeability (10.2 to 22.4 Barrer) and CO_2/CH_4 selectivity (33.6 to 36.4). As shown in Figure 9, the modification of the nanoparticles allowed for significantly improving the interface morphology between the nanoparticles and the polymer phase, preventing the formation of interfacial voids. Effect of surface modification was investigated also for zeolite NaY. Mixed matrix membrane were fabricated embedding the pristine and modified nanoparticles (loading range: 0–25 wt.%) in cellulose acetate [156]. Surface modification of zeolite with NH functional groups was performed in order to improve the CO_2 separation performances. However, in this case, the functionalization showed minor improvement when compared to the pristine nanoparticles (Table 7).

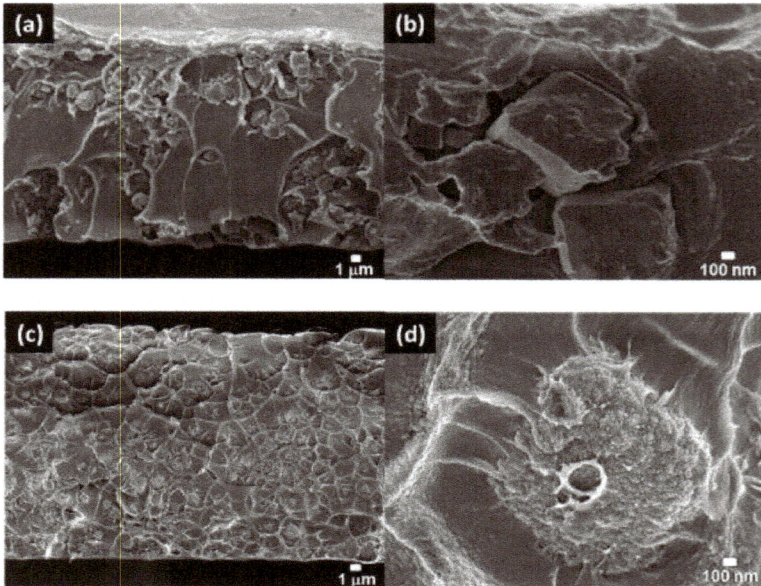

Figure 9. Cross section FESEM images of mixed-matrix membranes: (**a**,**b**) Matrimid with embedded pristine zeolite 5A; and, (**c**,**d**) Matrimid with embedded surface modified zeolite 5A [155], with copyright permission from © 2016 Elsevier.

Table 7. Gas separation performance of zeolites-based mixed matrix membranes (operating conditions ranging within 1–5 bar, 20–35 °C, unless differently specified).

Filler	Polymer	Loading (wt.%)	P_{CO_2} (Barrer)	α_{CO_2/N_2}	α_{CO_2/CH_4}	α_{CO_2/H_2}	Ref.
ZSM-5	PEBAX 1675	0	120		20.3		[143]
		5	230		21		
		10	191		32.5		
		15	170		33.9		
ZSM-5	Matrimid 5218	0	5.1 [a]		14.8		[144]
		6	6.6 [a]		15.6		
		15	11.1 [a]		7.2		
		24	14.5 [a]		4.8		
		30	21 [a]		3.6		
13X	PEBAX 1675	0	81.4	41			[145]
		10	104	39			
		15	114	47			
4A	PEBAX 1675	0	55.8	39.9	18.0		[147]
		5	71.4	51.0	32.5		
		10	97	53.9	26.2		
		20	113.7	39.2	17.5		
		30	155.8	13.0	7.9		

Table 7. *Cont.*

Filler	Polymer	Loading (wt.%)	P_{CO_2} (Barrer)	$\alpha_{CO2/N2}$	$\alpha_{CO2/CH4}$	$\alpha_{CO2/H2}$	Ref.
	PEBAX 1675	0	110	54	16	8.99	[148]
SAPO-34		9	100	53	16.5	8.29	
		23	130	56	21.9	6.58	
		33	250	55.7	16.4	8.96	
		50	340	55.5	16.5	8.40	
	PDMS	0	4796		3.0	4.21	[150]
4A		10	4226		2.7	1.55	
		20	3691		2.6	0.61	
		30	3323		2.9	0.40	
		40	2972		2.8	0.30	
		50	2886		2.9	0.27	
	PVAc	0	2.74	28	53		[151]
Ferrierite		20	3.93	61	54		
		40	3.93	82	57		
4A		20	2.55	52			
		40	2.73	74			
5A		20	2.77	46			
		40	1.70	33			
Silicalite-1		20	3.38	42			
		40	3.52	50			
	PTMSP	0	17430	0.9			[152]
Zeolite A		5	13029	9.7			
		20	11403	76.4			
ITQ-29		5	16501	4.4			
		20	14546	1.1			
	PSF	0	4.9		18.5		[154]
4A		20	5		12.5		
		25	6.9		7.6		
		30	7		2		
		35	7.12		1.44		
treated 4A		20	4.75		23.5		
		25	4.73		28		
		30	4.7		31		
		35	3.7		29		
	Matrimid	0	10.2		33.6		[155]
5A		10	26.7		31.3		
		20	31		30.8		
5A-Mg(OH)2		10	19.6		35.4		
		20	22.4		36.4		
	Cellulose Acetate	0	2.2	26			[156]
Na-Y		5	2.5	22.5			
		10	2.6	22			
		15	3.4	21			
		20	4.95	22.5			
		25	3.5	15			
Na-Y-NH2		5	3.2	25			
		10	3.5	23			
		15	3.65	22			
		20	4.1	26			
		25	4.3	17			

[a] Permeance (GPU); membrane thickness 3–5 μm.

According to the data reviewed, the fabrication of hybrid membranes containing zeolites can be promising for CO_2 capture applications. Loading up to 50 wt.% have been investigated, and rubbery materials (e.g., PEBAX or PDMS) showed good compatibility with the pristine nanoparticles, independently from their nature. Similar to MOFs, increasing the loading of pristine zeolites within polymeric phases enhances the CO_2 permeability of the hybrid membranes. However, this effect is mainly observed for low permeable polymers, since when high free volume polymers are used the hybrid membranes showed lower permeability when compared to the polymeric precursor. The effect on the selective features depends on the organic-inorganic interface, but a sieving effect for CO_2 is rarely observed. As observed for MOFs, surface modification is a suitable approach to improve the polymer-particles interface, but typically, the better compatibility mainly improves the selective features, and CO_2 permeability appears to be negligibly affected by the presence of the inorganic phase.

5. Porous Nanosheets

Two-dimensional nanoporous nanomaterials have been of great interest owing to their layered structure, which can significantly improve the sieving effect of nanoporous materials to gas transport. Inorganic membranes that are made of 2D metal organic frameworks have been reported in literature, showing promising separation performance [157,158]. 2D structures have also been reported for zeolites [159], and inorganic membranes have been fabricated [160,161], even though their potential for gas separation remains unexplored. The high aspect ratio of two-dimensional nanoporous particles makes them extremely attractive for the fabrication of mixed matrix membranes. Layered fillers perpendicular to the concentration gradient of the gas species in the membranes can give rise to outstanding separation performance because of a significant increase in tortuosity, hence in diffusive pathways, of the penetrants that cannot permeate through the nanoporous structure (Figure 10). A comparison in the water/ethanol separation performance of ZIF-8 and its 2D derivate (ZIF-L) showed a simultaneous improvement of both permeability and selectivity at even lower MOF loading [162]. Next generation of hybrid membranes containing porous nanosheets that are incorporated in polymer matrix will provide a solution in order to enhance the separation performance of membranes for CO_2 separation.

Porous layered and delaminated materials, with an intermediate structure between clay-like morphology and porous frameworks, represent an interesting class of porous 2D nanofillers. These materials can be exfoliated from bulk crystals, giving rise to high aspect ratio structures containing a porous architecture that can be of interest for gas separation applications. Layered aluminophospates (AlPO), layered silicates (AMH-3), layered titanosilicates JDF-L1, and layered COFs (NUS-2/3) are some examples that have been used for the fabrication of mixed matrix membranes. Nevertheless, very few studies have been dedicated to CO_2 separation, since selective sieving of H_2 from bigger molecules like CH_4 have been investigated to a bigger extent.

A pioneering work was developed by Kim et al. [163], where nanoporous layered silicate AMH-3 (pore size 3.4 Å) was first exfoliated and subsequently embedded in cellulose acetate, achieving a loading up to 6 wt.%. The CO_2 permeability increased along with the inorganic loading, and this enhancement was attributed to the competing effects of transport through the nanopores, the interlayer spaces, and through a lower-density cellulose acetate phase. Negligible influence was observed on the selective features. Kang et al. [164] reported novel synthesis of NUS-2 and NUS-3 layered materials that are based on COFs with excellent water and acid stability. Both of the COFs exhibit hexagonal channels with diameters of 0.8 nm and 1.8 nm for NUS-2 and NUS-3, respectively. The flower-like nanofillers contain leafs of 1–2 μm length and 50–100 nm width. The synthesized nanofillers were dispersed in two different polymer matrices Ultem (PEI) and polybenzimidazole (PBI) and separation performances for H_2/CO_2 and CO_2/CH_4 were studied. For CO_2/CH_4 separation with Ultem, both of the nanofillers increased CO_2 permeability and selectivity at 10 and 20 wt.% loading moving the pristine polymers closer to the upper bound. However, when the filler content was increased to 30 wt.%, the selective features of the membrane dropped, possibly due to void formation at the polymer/filler interface. On the other hand, for H_2 separation from CO_2, the PBI sample containing 20 wt.% NUS-2 surpassed the upper bound thanks to an impressive enhancement of the H_2/CO_2 selectivity. Alternatively, NUS-3 increased the permeability while maintaining or decreasing the selectivity. The highest H_2 permeability was obtained at 30 wt.% loading, which is 17 times the permeability of pristine PBI, which is followed by a 50% reduction in selectivity.

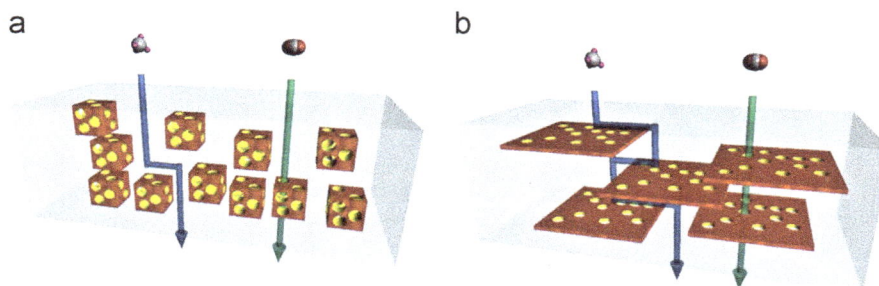

Figure 10. Schematic representation of the effect of isotropic particles (**a**) and nanoporous sheets (**b**) on the transport through mixed matrix membranes. Reprinted from [163], with copyright permission from © 2013 Elsevier.

Rodenas et al. [165], in 2014, synthesized and compared CuBDC MOFs with three different morphologies: isotropic nanocrystals (nc-CuBDC), bulk-type crystals (b-CuBDC), and nanosheets (ns-CuBDC). The different nanoparticles were embedded within a polyimide-based (Matrimid 5218) polymeric matrix. It was shown that the CuBDC offered large surface area, which was about five-fold higher than the one that was obtained for the b-CuBDC. Mixed gas permeation tests showed that the addition of both nc-CuBDC and b-CuBDC (8 wt.% loading) determined a drop in the CO_2/CH_4 selectivity when compared to the pristine polymer. However, when a similar loading of ns-CuBDC was embedded in the polymeric matrix, a 30% enhancement in the separation factor was observed. This effect was even more evident when the feed pressure was increased from 3 to 7.5 bar, where the selectivity improvement reached a 80% higher value as compared to the pristine polymer. At 3 bar feed pressure, the CO_2 permeability gradually increased from 5.78 to 9.91 Barrer (at 3.7 wt.% loading) and then decreased to 4.09 Barrer (at 8.3 wt.%, Table 8). At a similar loading of bulk and nanocrystals, a minor reduction in CO_2 permeability was observed. Interestingly, the embedment of ns-CuBDC was also reported to limit the effect of CO_2-induced plasticization characteristic of polyimides at high partial pressure of CO_2. The authors attributed this effect to the depletion of MOF-free permeation pathways, sustaining the selective features of the membrane under high CO_2 concentration within the hybrid matrix. A similar work has also been recently reported by Shete et al. [166], who embedded Cu-based MOF nanosheets (lateral size 2.5 μm, thickness 25 nm) in Matrimid 5218. Results that were obtained by the two studies are quite similar, with a decrease in CO_2 permeability at increasing the nanosheets loading with improved selectivity (Table 8), strengthening the consistency of the influence of nanosheets on the transport properties of polyimides. The main difference is related to the influence of the feed pressure: in the latter case, the mixed matrix membranes selectivity decreased with increasing the operating pressure, whereas an opposite trend was observed in the other study.

The CO_2/CH_4 gas separation performance of ultrathin layer that was obtained by dispersing 2D MOFs in PIM-1 was investigated by Cheng et al. [167]. CuBDC nanosheets with a thickness of 15 to 40 nm and ~100 aspect ratio were successfully embedded into PIM-1 up to 5 wt.% loading. Thin films (200 to 2200 nm) were then coated on a porous Al_2O_3 support via spin-coating technique. At 10% loading, the enhancement in CO_2 selectivity from 4.4 to 16 (~300% increase) was observed. Nevertheless, the selectivity improvement with an increase in MOF loading was at significant expense of the CO_2 permeance, which dropped from 1750 to 500 GPU with a 2 wt.% loading. Interestingly, no significant differences in permeance have been observed between 2, 5, 10, and 15 wt.% loading, suggesting that the transport is dominated by the embedded phase already in the low loadings (Table 8). The gas selectivity enhancement was attributed to the tortuosity and the pathway created by centrifugal force, which helped to align nanosheets horizontally. At higher loading up to 15 wt.%, the selectivity reduction was observed mainly due to the presence of nonselective voids and agglomeration.

Yang et al. [168] recently reported the influence of CuBDC nanosheets on the performance of high free volume polymers, such as PIM-1 and 6FDA-DAM. Nanosheets with a lateral dimension of 1–8 μm and a thickness of 40 nm were synthesized and embedded in the polymer phase via sonication. As observed previously for PIM-1, the incorporation of nanosheets resulted in a decrease of CO_2 permeability at the low loadings for both PIM-1 and 6FDA-DAM (Table 8). Notably, in the case of PIM-1 small differences were observed between the two different filler loadings, whereas in the case of 6FDA-DAM, the permeability decrease was more evident between the 2 and 4 wt.% loading. In both cases, the presence of the porous nanosheets led to a significant increase (20–40%) in the selective feature of the hybrid membranes.

In another study, Kang et al. [169] prepared MMMs with a newly synthesized 2D MOF (10×100 nm^2), [Cu$_2$(ndc)$_2$(dabco)]n, (ndc = 1,4-naphthalene dicarboxylate, dabco = 1,4-diazabicyclo[2.2.2]octane), and incorporated into PBI (polybenzimidazole) matrix for pre-combustion CO_2 separation. MOF loading from 10 to 20 wt.% provided highly selective MMMs, with about 100% increment in H_2/CO_2 ideal selectivity. The authors attributed the high selective features to fast H_2 permeation through the MOF, whereas CO_2 follows slower diffusive pathways in view of the larger kinetic diameter. Higher MOF loadings (> 20 wt.%) resulted in a selectivity drop in selectivity, which is possibly due to void formation. Comparison of different morphologies showed that MOF nanosheets offered better selectivity and permeability of the hybrid membranes because of the shape, orientation, and interfacial adhesion inside the matrix. As in the previous case, similar loadings of bulk or nanocrystals (20 wt.%) showed lower selectivity values compared to the nanosheet morphology.

Table 8. Gas separation performance of mixed matrix membranes containing MOFs nanosheets (operating conditions ranging within 1–5 bar, 20–35 °C, mixed gas conditions unless differently specified).

Filler	Polymer	Loading (wt.%)	P_{CO2} (Barrer)	$\alpha_{CO2/N2}$	$\alpha_{CO2/CH4}$	$\alpha_{H2/CO2}$	Ref.
	Cellulose Acetate	0	7.55		29.61		[163]
AMH-3		2	9.65		29.24		
		4	10.36		30.03		
		6	11.59		29.71		
	Ultem	0 [a]	2.22		20.2	2.88	[164]
NUS-2		10 [a]	3.75		25	3.39	
		20 [a]	4.92		22.4	4.61	
		30 [a]	8.70		12.7	1.89	
NUS-3		10 [a]	5.89		22.7	2.46	
		20 [a]	15		28.3	2.23	
		30 [a]	8.11		10.7	2.45	
	Matrimid 5218	0	5.78		59.8		[165]
ns-CuBDC [b]		1.7	5.38		61.6		
		3.7	9.91		59.5		
		4.3	4.74		63.5		
		8.2	4.09		78.7		
b-CuBDC [b]		7.9	5.21		45		
nc-CuBDC [b]		8.3	5.03		49.4		
	Matrimid 5218	0	7.2 [c]	23.7			[166]
CuBDC		4	6.4 [c]	42.0			
		8	4.0 [c]	48.1			
		0	15.2		25.3		
		12	6.6		40.3		

Table 8. *Cont.*

Filler	Polymer	Loading (wt.%)	P_{CO_2} (Barrer)	α CO2/N2	α CO2/CH4	α H2/CO2	Ref.
	PIM-1	0	1750 [d]		4.4		[167]
CuBDC-ns		2	500 [d]		10.2		
		5	490 [d]		12.9		
		10	400 [d]		16.0		
		15	490 [d]		11.7		
		0	161 [d]		12.2		
		10	196 [d]		10.8		
		10	407 [d]		15.5		
	PIM-1	0	3100		17		[168]
CuBDC-ns		2	2030		24		
		4	2300		22		
	6FDA-DAM	0	590		30		
CuBDC-ns		2	570		37		
		4	430		43		
	PBI	0	3.62			9.3	[169]
ns-Cu$_2$(ndc)$_2$(dabco) [b]		10	4.86			18.7	
		20	6.15			22.8	
		30	11.9			12.3	
		50	66.4			4.8	
bc-Cu$_2$(ndc)$_2$(dabco) [b]		20	5.18			12.6	
nc-Cu$_2$(ndc)$_2$(dabco) [b]		20	5.29			17.6	

[a] Operating pressure of 2 bar, [b] ns = nanosheets; bc = bulk crystals; nc = nano crystals; [c] single gas tests; [d] permeance (GPU).

Despite the early stage of the research, the analysis of the performance achieved while using 2D nanoporous materials for the fabrication of mixed matrix membranes clearly showed a promising potential within CO_2 capture. Systematically, the 2D shape was demonstrated to be able to achieve better performance when compared to the isotropic particles, independently from their size. Interestingly, compared to isotropic nanoparticles, the effect of nanosheets is already visible in the low loading range, similar to what has been observed for graphene [22]. The use of 2D porous nanoparticles can be of particular interest for the enhancement of the selective feature of high free volume polymers, where a partial loss in CO_2 permeability can be tolerated if being counterbalanced by a significant enhancement of the separation factor. A notable increase of studies that are dedicated to this topic is expected in the near future.

6. Conclusions and Perspective

The recent advances in the synthesis and improvements of 2D and 3D porous nanophases has driven a continuous research within the development of mixed matrix membranes for gas separation purposes. In particular, the possibility of tuning the pore diameter to a gas-sieving level and the CO_2-philicity of the pore cavity has the potential to facilitate the simultaneous enhancement of the solubility and diffusivity coefficient of carbon dioxide. Therefore, CO_2 permeability and selectivity can be expected to benefit from these features, leading to a shift in the separation performance towards the upper right corner of the Robeson plot.

Notable attention has been given to MOF nanoparticles and MOFs nanosheets. The pore opening size falling within the gas kinetic diameters and the presence of unsaturated open metal sites makes them of particular interest for CO_2 separation. Analysis of adding ZIF nanoparticles into highly or moderately permeable polymeric membrane materials reveals a clear tendency to improve the CO_2 permeability when the nanofiller loading is increased to 30–40 wt.%. The incorporation of ZIFs has been frequently reported to be associated to the disruption the polymer chain packing, leading to an increase of the MMMs free volume. However, selectivity enhancement was seldom reported despite the expected sieving effect and the observed suitable interface morphology. The ZIFs flexible framework is expected to be among the main reasons for this phenomenon. Also, in the case of other analyzed MOFs (UiO-66, MILs and various metallic-based MOFs), the CO_2 permeability enhancement was frequently observed, with the enhancement being proportional to the MOF content. However, the increase in selective feature was typically reported only at low particles loading (especially for UiO-66) and mild operative conditions, suggesting that the sieving ability of the pore opening is not extremely effective for gas separation purpose. Structural flexibility and poor interface interactions were frequently mentioned as the possible causes. Therefore, the achievement of a more rigid structure of the MOFs cage and more effect functionalization are desirable to improve the efficiency of the embedded phase. Interestingly, particles with smaller size have shown to be more effective compared to inorganic phases with bigger size. In addition, particles with reduced size can facilitate the fabrication of thin (<1 μm) selective layers.

Porous nanosheets showed a promising potential for the fabrication of mixed matrix membranes for CO_2 separation. When compared to 3D porous materials, the impact of 2D nonporous materials is demonstrated even at low loading range (<10 wt.%). The use of 2D shape was systematically demonstrated to obtain better performance compared to isotropic particles. Higher selectivity can be achieved using MOF nanosheets, even when they are incorporated in high free volume polymers, but the variation typically takes place to the expense of the gas transport through the selective layer. The intrinsic nature of these 2D nanoparticles has the potential to be a successful strategy to efficiently fabricate mixed matrix membranes with superior separation performance in the form of thin composite membranes. Therefore, future work has to focus on the reduction of the thickness of these 2D porous layers, allowing for achieving membrane thickness in the order of 100–200 nm.

Porous organic frameworks (POFs) have also been recently investigated for CO_2 separation. Their fully-organic nature facilitates their dispersion in polymer phases, but their more rigid structure confers interesting feature to the hybrid membranes. Experimental results gave evidence of an unprecedented capacity of stopping physical aging in high free volume polymers. Even though CO_2 permeability is frequently enhanced using PAFs, negligible influence on selectivity of the hybrids was observed. Nevertheless, their promising performance has been disclosed only for thick self-standing membranes, and more investigation on their efficiency for thin films are needed to fully understand their potential.

Zeolites, as one of the most common fillers, attracted great interest in MMMs fabrication and have been investigated for last two decades. When compared to MOFs, the absence of organic ligand in the lattice, the control of zeolite/polymer interface is more difficult than MOF/polymer interface. Therefore, many efforts have been spent to ensure the achievement of proper interface morphology to reduce the negative effects that are associated to interfacial voids. Loading of up to 50 wt.% has been reported, and rubbery polymers (e.g., PDMS) showed good compatibility and adhesion with pristine nanoparticles. Increase in zeolite content led to higher permeability and effect of pristine zeolites on CO_2 permeability was more pronounced for low permeable polymers when compared to high free volume polymers. Surface modification of zeolites have shown better compatibility and improved selectivity with negligible effect in CO_2 permeability.

The following focuses may be of appreciable impact in the future development of MMMs with superior transport properties:

- to reduce primary particle size of existing MOFs and expedite their incorporation in thin composite polymeric membranes;
- to increase the CO_2 affinity and the polymer/particle interactions by novel surface functionalization procedures on the nanoparticle or by post-functionalization after membrane preparation, aiming at improving the CO_2 separation performance and simplifying their dispersion in the polymeric phases;
- to tune the structure and morphology of POFs with the aim of enhancing the selectivity of hybrid matrix when used in high free volume polymers;

- to design and fabricate novel 2D MOF frameworks with improved sieving ability that do not sacrifice the gas transport through the selective layer; and,
- to systematically investigate the potential of hybrid membranes in H_2 separation, exploiting the exceptional H_2 sieving ability of some MOFs.

Funding: "This research and the APC were funded by the European Union's Horizon 2020 Research and Innovation program, grant number 727734".

Conflicts of Interest: The authors declare no conflict of interest.

References

1. Pera-Titus, M. Porous Inorganic Membranes for CO2 Capture: Present and Prospects. *Chem. Rev.* **2014**, *114*, 1413–1492. [CrossRef] [PubMed]
2. Rackley, S.A. Chapter 4—Carbon Capture from Power Generation. In *Carbon Capture and Storage*; Rackley, S.A., Ed.; Butterworth-Heinemann: Oxford, UK, 2010; pp. 65–93, ISBN 978-1-85617-636-1.
3. Rackley, S.A. Chapter 5—Carbon Capture from Industrial Processes. In *Carbon Capture and Storage*; Rackley, S.A., Ed.; Butterworth-Heinemann: Oxford, UK, 2010; pp. 95–102, ISBN 978-1-85617-636-1.
4. Boot-Handford, M.E.; Abanades, J.C.; Anthony, E.J.; Blunt, M.J.; Brandani, S.; Mac Dowell, N.; Fernandez, J.R.; Ferrari, M.C.; Gross, R.; Hallett, J.P.; et al. Carbon capture and storage update. *Energy Environ. Sci.* **2014**, *7*, 130–189. [CrossRef]
5. Dalane, K.; Dai, Z.; Mogseth, G.; Hillestad, M.; Deng, L. Potential applications of membrane separation for subsea natural gas processing: A review. *J. Nat. Gas Sci. Eng.* **2017**, *39*, 101–117. [CrossRef]
6. Ahmadi, M.; Tas, E.; Kilic, A.; Kumbaraci, V.; Talinli, N.; Ahunbay, M.G.; Tantekin-Ersolmaz, S.B. Highly CO_2 Selective Microporous Metal-Imidazolate Framework (MMIF) Based Mixed Matrix Membranes. *ACS Appl. Mater. Interfaces* **2017**, *9*, 35936–35946. [CrossRef] [PubMed]
7. Yu, J.; Xie, L.H.; Li, J.R.; Ma, Y.; Seminario, J.M.; Balbuena, P.B. CO_2 Capture and Separations Using MOFs: Computational and Experimental Studies. *Chem. Rev.* **2017**, *117*, 9674–9754. [CrossRef] [PubMed]
8. Koros, W.J.; Zhang, C. Materials for next-generation molecularly selective synthetic membranes. *Nat. Mater.* **2017**, *16*, 289–297. [CrossRef] [PubMed]
9. Dai, Z.; Ansaloni, L.; Deng, L. Recent advances in multi-layer composite polymeric membranes for CO_2 separation: A review. *Green Energy Environ.* **2012**, *1*, 102–128. [CrossRef]
10. Ansaloni, L.; Deng, L. Advances in polymer-inorganic hybrids as membrane materials. In *Recent Developments in Polymer Macro, Micro and Nano Blends: Preparation and Characterisation*; Woodhead Publishing: Sawston, UK; Cambridge, UK, 2016; pp. 163–206. ISBN 9780081004272.
11. Robeson, L.M. Correlation of separation factor versus permeability for polymeric membranes. *J. Memb. Sci.* **1991**, *62*, 165–185. [CrossRef]
12. Robeson, L.M. The upper bound revisited. *J. Membr. Sci.* **2008**, *320*, 390–400. [CrossRef]
13. Freeman, B.D. Basis of Permeability/Selectivity Tradeoff Relations in Polymeric Gas Separation Membranes. *Macromolecules* **1999**, *32*, 375–380. [CrossRef]
14. Robeson, L.M.; Smith, Z.P.; Freeman, B.D.; Paul, D.R. Contributions of diffusion and solubility selectivity to the upper bound analysis for glassy gas separation membranes. *J. Membr. Sci.* **2014**, *453*, 71–83. [CrossRef]
15. Park, H.B.; Jung, C.H.; Lee, Y.M.; Hill, A.J.; Pas, S.J.; Mudie, S.T.; Van Wagner, E.; Freeman, B.D.; Cookson, D.J. Polymers with cavities tuned for fast selective transport of small molecules and ions. *Science* **2007**, *318*, 254–258. [CrossRef] [PubMed]
16. Low, Z.-X.; Budd, P.M.; McKeown, N.B.; Patterson, D.A. Gas Permeation Properties, Physical Aging, and Its Mitigation in High Free Volume Glassy Polymers. *Chem. Rev.* **2018**. [CrossRef] [PubMed]
17. Rafiq, S.; Deng, L.; Hägg, M.-B. Role of Facilitated Transport Membranes and Composite Membranes for Efficient CO_2 Capture—A Review. *ChemBioEng Rev.* **2016**, *3*, 68–85. [CrossRef]
18. Caro, J. Hierarchy in inorganic membranes. *Chem. Soc. Rev.* **2016**, *45*, 3468–3478. [CrossRef] [PubMed]
19. Adams, R.; Carson, C.; Ward, J.; Tannenbaum, R.; Koros, W. Metal organic framework mixed matrix membranes for gas separations. *Microporous Mesoporous Mater.* **2010**, *131*, 13–20. [CrossRef]

20. Galizia, M.; Chi, W.S.; Smith, Z.P.; Merkel, T.C.; Baker, R.W.; Freeman, B.D. 50th Anniversary Perspective: Polymers and Mixed Matrix Membranes for Gas and Vapor Separation: A Review and Prospective Opportunities. *Macromolecules* **2017**, *50*, 7809–7843. [CrossRef]

21. Dong, G.; Li, H.; Chen, V. Challenges and opportunities for mixed-matrix membranes for gas separation. *J. Mater. Chem. A* **2013**, *1*, 4610–4630. [CrossRef]

22. Janakiram, S.; Ahmadi, M.; Dai, Z.; Ansaloni, L.; Deng, L. Performance of nanocomposite membranes containing 0D to 2D nanofillers for CO_2 separation: A review. *Membranes* **2018**, *8*, 24. [CrossRef] [PubMed]

23. Rezakazemi, M.; Ebadi Amooghin, A.; Montazer-Rahmati, M.M.; Ismail, A.F.; Matsuura, T. State-of-the-art membrane based CO_2 separation using mixed matrix membranes (MMMs): An overview on current status and future directions. *Prog. Polym. Sci.* **2014**, *39*, 817–861. [CrossRef]

24. Seoane, B.; Coronas, J.; Gascon, I.; Benavides, M.E.; Karvan, O.; Caro, J.; Kapteijn, F.; Gascon, J. Metal–organic framework based mixed matrix membranes: A solution for highly efficient CO_2 capture? *Chem. Soc. Rev.* **2015**, *44*, 2421–2454. [CrossRef] [PubMed]

25. Jusoh, N.; Fong Yeong, Y.; Leng Chew, T.; Keong Lau, K.; Mohd Shariff, A. Separation & Purification Reviews Current Development and Challenges of Mixed Matrix Membranes for CO_2/CH_4 Separation Current Development and Challenges of Mixed Matrix Membranes for CO_2/CH_4 Separation. *Sep. Purif. Rev.* **2016**, *45*, 321–344. [CrossRef]

26. Vinoba, M.; Bhagiyalakshmi, M.; Alqaheem, Y.; Alomair, A.A.; Pérez, A.; Rana, M.S. Recent progress of fillers in mixed matrix membranes for CO_2 separation: A review. *Sep. Purif. Technol.* **2017**, *188*, 431–450. [CrossRef]

27. Wang, M.; Wang, Z.; Zhao, S.; Wang, J.; Wang, S. Recent advances on mixed matrix membranes for CO2 separation. *Chinese J. Chem. Eng.* **2017**, *25*, 1581–1597. [CrossRef]

28. Zhou, H.-C.; Long, J.R.; Yaghi, O.M. Introduction to Metal–Organic Frameworks. *Chem. Rev.* **2012**, *112*, 673–674. [CrossRef] [PubMed]

29. Farrusseng, D.; Aguado, S.; Pinel, C. Metal-Organic Frameworks: Opportunities for Catalysis. *Angew. Chemie Int. Ed.* **2009**, *48*, 7502–7513. [CrossRef] [PubMed]

30. Chen, B.; Yang, Z.; Zhu, Y.; Xia, Y. Zeolitic imidazolate framework materials: recent progress in synthesis and applications. *J. Mater. Chem. A* **2014**, *2*, 16811–16831. [CrossRef]

31. James, S.L. Metal-organic frameworks. *Chem. Soc. Rev.* **2003**, *32*, 276–288. [CrossRef] [PubMed]

32. Stock, N.; Biswas, S. Synthesis of metal-organic frameworks (MOFs): routes to various MOF topologies, morphologies, and composites. *Chem. Rev.* **2011**, *112*, 933–969. [CrossRef] [PubMed]

33. Hyun, S.; Hwa Lee, J.; Yeong Jung, G.; Kyeong Kim, Y.; Kyung Kim, T.; Jeoung, S.; Kyu Kwak, S.; Moon, D.; Ri Moon, H. Exploration of Gate-Opening and Breathing Phenomena in a Tailored Flexible Metal—Organic Framework. *Inorg. Chem* **2016**, *55*, 1920–1925. [CrossRef] [PubMed]

34. Schneemann, A.; Bon, V.; Schwedler, I.; Senkovska, I.; Kaskel, S.; Fischer, R.A. Flexible metal–organic frameworks. *Chem. Soc. Rev.* **2014**, *43*, 6062–6096. [CrossRef] [PubMed]

35. Castellanos, S.; Kapteijn, F.; Gascon, J. Photoswitchable metal organic frameworks: turn on the lights and close the windows. *CrystEngComm* **2016**, *18*, 4006–4012. [CrossRef]

36. Alhamami, M.; Doan, H.; Cheng, C.-H. A Review on Breathing Behaviors of Metal-Organic-Frameworks (MOFs) for Gas Adsorption. *Materials (Basel)* **2014**, *7*, 3198–3250. [CrossRef] [PubMed]

37. Fairen-Jimenez, D.; Moggach, S.A.; Wharmby, M.T.; Wright, P.A.; Parsons, S.; Duren, T. Opening the gate: framework flexibility in ZIF-8 explored by experiments and simulations. *J. Am. Chem. Soc.* **2011**, *133*, 8900–8902. [CrossRef] [PubMed]

38. Kolokolov, D.I.; Maryasov, A.G.; Ollivier, J.; Freude, D.; Haase, J.; Stepanov, A.G.; Jobic, H. Uncovering the Rotation and Translational Mobility of Benzene Confined in UiO-66 (Zr) Metal–Organic Framework by the ^{2}H NMR–QENS Experimental Toolbox. *J. Phys. Chem. C* **2017**, *121*, 2844–2857. [CrossRef]

39. Damron, J.T.; Ma, J.; Kurz, R.; Saalwächter, K.; Matzger, A.J.; Ramamoorthy, A. The Influence of Chemical Modification on Linker Rotational Dynamics in Metal-Organic Frameworks. *Angew. Chem. Int. Ed.* **2018**, *57*, 8678–8681. [CrossRef] [PubMed]

40. Yoo, G.Y.; Lee, W.R.; Jo, H.; Park, J.; Song, J.H.; Lim, K.S.; Moon, D.; Jung, H.; Lim, J.; Han, S.S.; Jung, Y.; Hong, C.S. Adsorption of Carbon Dioxide on Unsaturated Metal Sites in M_2 (dobpdc) Frameworks with Exceptional Structural Stability and Relation between Lewis Acidity and Adsorption Enthalpy. *Chem. - A Eur. J.* **2016**, *22*, 7444–7451. [CrossRef] [PubMed]

41. Poloni, R.; Lee, K.; Berger, R.F.; Smit, B.; Neaton, J.B. Understanding Trends in CO_2 Adsorption in Metal−Organic Frameworks with Open-Metal Sites. *J. Phys. Chem. Lett.* **2014**, *5*, 861–865. [CrossRef] [PubMed]

42. Thornton, A.W.; Dubbeldam, D.; Liu, M.S.; Ladewig, B.P.; Hill, A.J.; Hill, M.R. Feasibility of zeolitic imidazolate framework membranes for clean energy applications. *Energy Environ. Sci.* **2012**, *5*, 7637–7646. [CrossRef]

43. Bhattacharjee, S.; Jang, M.-S.; Kwon, H.-J.; Ahn, W.-S. Zeolitic imidazolate frameworks: Synthesis, functionalization, and catalytic/adsorption applications. *Catal. Surv. Asia* **2014**, *18*, 101–127. [CrossRef]

44. Park, K.S.; Ni, Z.; Côté, A.P.; Choi, J.Y.; Huang, R.; Uribe-Romo, F.J.; Chae, H.K.; O'Keeffe, M.; Yaghi, O.M. Exceptional chemical and thermal stability of zeolitic imidazolate frameworks. *Proc. Natl. Acad. Sci. USA* **2006**, *103*, 10186–10191. [CrossRef] [PubMed]

45. Banerjee, R.; Phan, A.; Wang, B.; Knobler, C.; Furukawa, H.; O'keeffe, M.; Yaghi, O.M. High-throughput synthesis of zeolitic imidazolate frameworks and application to CO_2 capture. *Science* **2008**, *319*, 939–943. [CrossRef] [PubMed]

46. Fairen-Jimenez, D.; Galvelis, R.; Torrisi, A.; Gellan, A.D.; Wharmby, M.T.; Wright, P.A.; Mellot-Draznieks, C.; Düren, T. Flexibility and swing effect on the adsorption of energy-related gases on ZIF-8: combined experimental and simulation study. *Dalt. Trans.* **2012**, *41*, 10752–10762. [CrossRef] [PubMed]

47. Coudert, F.-X. Molecular Mechanism of Swing Effect in Zeolitic Imidazolate Framework ZIF-8: Continuous Deformation upon Adsorption. *ChemPhysChem* **2017**, *18*, 2732–2738. [CrossRef] [PubMed]

48. Ordoñez, M.J.C.; Balkus, K.J., Jr.; Ferraris, J.P.; Musselman, I.H. Molecular sieving realized with ZIF-8/Matrimid®mixed-matrix membranes. *J. Memb. Sci.* **2010**, *361*, 28–37. [CrossRef]

49. Basu, S.; Cano-Odena, A.; Vankelecom, I.F.J. MOF-containing mixed-matrix membranes for CO_2/CH_4 and CO_2/N_2 binary gas mixture separations. *Sep. Purif. Technol.* **2011**, *81*, 31–40. [CrossRef]

50. Song, Q.; Nataraj, S.K.; Roussenova, M.V; Tan, J.C.; Hughes, D.J.; Li, W.; Bourgoin, P.; Alam, M.A.; Cheetham, A.K.; Al-Muhtaseb, S.A.; et al. Zeolitic imidazolate framework (ZIF-8) based polymer nanocomposite membranes for gas separation. *Energy Environ. Sci.* **2012**, *5*, 8359–8369. [CrossRef]

51. Thompson, J.A.; Chapman, K.W.; Koros, W.J.; Jones, C.W.; Nair, S. Sonication-induced Ostwald ripening of ZIF-8 nanoparticles and formation of ZIF-8/polymer composite membranes. *Microporous Mesoporous Mater.* **2012**, *158*, 292–299. [CrossRef]

52. Thompson, J.A.; Vaughn, J.T.; Brunelli, N.A.; Koros, W.J.; Jones, C.W.; Nair, S. Mixed-linker zeolitic imidazolate framework mixed-matrix membranes for aggressive CO_2 separation from natural gas. *Microporous Mesoporous Mater.* **2014**, *192*, 43–51. [CrossRef]

53. Casado-Coterillo, C.; Fernandez-Barquin, A.; Zornoza, B.; Tellez, C.; Coronas, J.; Irabien, A. Synthesis and characterisation of MOF/ionic liquid/chitosan mixed matrix membranes for CO_2/N_2 separation. *RSC Adv.* **2015**, *5*, 102350–102361. [CrossRef]

54. Carter, D.; Tezel, F.H.; Kruczek, B.; Kalipcilar, H. Investigation and comparison of mixed matrix membranes composed of polyimide matrimid with ZIF-8, silicalite, and SAPO-34. *J. Memb. Sci.* **2017**, *544*, 35–46. [CrossRef]

55. Guo, A.; Ban, Y.; Yang, K.; Yang, W. Metal-organic framework-based mixed matrix membranes: Synergetic effect of adsorption and diffusion for CO_2/CH_4 separation. *J. Memb. Sci.* **2018**, *562*, 76–84. [CrossRef]

56. Jusoh, N.; Yeong, Y.F.; Lau, K.K.; Shariff, A.M. Transport properties of mixed matrix membranes encompassing zeolitic imidazolate framework 8 (ZIF-8) nanofiller and 6FDA-durene polymer: Optimization of process variables for the separation of CO_2 from CH_4. *J. Clean. Prod.* **2017**, *149*, 80–95. [CrossRef]

57. Wijenayake, S.N.; Panapitiya, N.P.; Versteeg, S.H.; Nguyen, C.N.; Goel, S.; Balkus, K.J.; Musselman, I.H.; Ferraris, J.P. Surface cross-linking of ZIF-8/polyimide mixed matrix membranes (MMMs) for gas separation. *Ind. Eng. Chem. Res.* **2013**, *52*, 6991–7001. [CrossRef]

58. Zhang, Z.; Xian, S.; Xia, Q.; Wang, H.; Li, Z.; Li, J. Enhancement of CO_2 adsorption and CO_2/N_2 selectivity on ZIF-8 via postsynthetic modification. *AIChE J.* **2013**, *59*, 2195–2206. [CrossRef]

59. Askari, M.; Chung, T.-S. Natural gas purification and olefin/paraffin separation using thermal cross-linkable co-polyimide/ZIF-8 mixed matrix membranes. *J. Memb. Sci.* **2013**, *444*, 173–183. [CrossRef]

60. Nafisi, V.; Hägg, M.-B. Gas separation properties of ZIF-8/6FDA-durene diamine mixed matrix membrane. *Sep. Purif. Technol.* **2014**, *128*, 31–38. [CrossRef]

61. Nafisi, V.; Hägg, M.B. Development of dual layer of ZIF-8/PEBAX-2533 mixed matrix membrane for CO_2 capture. *J. Memb. Sci.* **2014**, *459*, 244–255. [CrossRef]

62. Sánchez-Laínez, J.; Zornoza, B.; Téllez, C.; Coronas, J. Asymmetric polybenzimidazole membranes with thin selective skin layer containing ZIF-8 for H_2/CO_2 separation at pre-combustion capture conditions. *J. Memb. Sci.* **2018**, *563*, 427–434. [CrossRef]

63. Dai, Y.; Johnson, J.R.; Karvan, O.; Sholl, D.S.; Koros, W.J. Ultem®/ZIF-8 mixed matrix hollow fiber membranes for CO_2/N_2 separations. *J. Memb. Sci.* **2012**, *401–402*, 76–82. [CrossRef]

64. Mubashir, M.; Fong, Y.Y.; Leng, C.T.; Keong, L.K. Issues and Current Trends of Hollow-Fiber Mixed-Matrix Membranes for CO_2 Separation from N_2 and CH_4. *Chem. Eng. Technol.* **2018**, *41*, 235–252. [CrossRef]

65. Zhu, J.; Qin, L.; Uliana, A.; Hou, J.; Wang, J.; Zhang, Y.; Li, X.; Yuan, S.; Li, J.; Tian, M.; et al. Elevated Performance of Thin Film Nanocomposite Membranes Enabled by Modified Hydrophilic MOFs for Nanofiltration. *ACS Appl. Mater. Interfaces* **2017**, *9*, 1975–1986. [CrossRef] [PubMed]

66. Sorribas, S.; Gorgojo, P.; Téllez, C.; Coronas, J.; Livingston, A.G. High Flux Thin Film Nanocomposite Membranes Based on Metal–Organic Frameworks for Organic Solvent Nanofiltration. *J. Am. Chem. Soc.* **2013**, *135*, 15201–15208. [CrossRef] [PubMed]

67. Xiao, F.; Wang, B.; Hu, X.; Nair, S.; Chen, Y. Thin film nanocomposite membrane containing zeolitic imidazolate framework-8 via interfacial polymerization for highly permeable nanofiltration. *J. Taiwan Inst. Chem. Eng.* **2018**, *83*, 159–167. [CrossRef]

68. Echaide-Górriz, C.; Navarro, M.; Téllez, C.; Coronas, J. Simultaneous use of MOFs MIL-101(Cr) and ZIF-11 in thin film nanocomposite membranes for organic solvent nanofiltration. *Dalt. Trans.* **2017**, *46*, 6244–6252. [CrossRef] [PubMed]

69. Sánchez-Laínez, J.; Paseta, L.; Navarro, M.; Zornoza, B.; Téllez, C.; Coronas, J. Ultrapermeable Thin Film ZIF-8/Polyamide Membrane for H_2/CO_2 Separation at High Temperature without Using Sweep Gas. *Adv. Mater. Interfaces* **2018**, 1800647. [CrossRef]

70. Li, Y.; Liang, F.; Bux, H.; Yang, W.; Caro, J. Zeolitic imidazolate framework ZIF-7 based molecular sieve membrane for hydrogen separation. *J. Memb. Sci.* **2010**, *354*, 48–54. [CrossRef]

71. Arami-Niya, A.; Birkett, G.; Zhu, Z.; Rufford, T.E. Gate opening effect of zeolitic imidazolate framework ZIF-7 for adsorption of CH_4 and CO_2 from N_2. *J. Mater. Chem. A* **2017**, *5*, 21389–21399. [CrossRef]

72. Li, T.; Pan, Y.; Peinemann, K.-V.; Lai, Z. Carbon dioxide selective mixed matrix composite membrane containing ZIF-7 nano-fillers. *J. Memb. Sci.* **2013**, *425*, 235–242. [CrossRef]

73. Al-maythalony, B.A.; Alloush, A.M.; Faizan, M.; Dafallah, H.; Elgzoly, M.A.A.; Seliman, A.A.A.; Al-ahmed, A.; Yamani, Z.H.; Habib, M.A.M.; Cordova, K.E.; et al. Tuning the Interplay between Selectivity and Permeability of ZIF-7 Mixed Matrix Membranes. *ACS Appl. Mater. Interfaces* **2017**, *9*, 33401–33407. [CrossRef] [PubMed]

74. Morris, W.; Doonan, C.J.; Furukawa, H.; Banerjee, R.; Yaghi, O.M. Crystals as molecules: postsynthesis covalent functionalization of zeolitic imidazolate frameworks. *J. Am. Chem. Soc.* **2008**, *130*, 12626–12627. [CrossRef] [PubMed]

75. Phan, A.; Doonan, C.J.; Uribe-Romo, F.J.; Knobler, C.B.; O'keeffe, M.; Yaghi, O.M. Synthesis, Structure, and Carbon Dioxide Capture Properties of Zeolitic Imidazolate Frameworks. *Acc. Chem. Res.* **2010**, *58*. [CrossRef] [PubMed]

76. Japip, S.; Xiao, Y.; Chung, T.-S. Particle-Size Effects on Gas Transport Properties of 6FDA-Durene/ZIF-71 Mixed Matrix Membranes. *Ind. Eng. Chem. Res.* **2016**, *55*, 9507–9517. [CrossRef]

77. Ehsani, A.; Pakizeh, M. Synthesis, characterization and gas permeation study of ZIF-11/Pebax® 2533 mixed matrix membranes. *J. Taiwan Inst. Chem. Eng.* **2016**, *66*, 414–423. [CrossRef]

78. Boroglu, M.S.; Yumru, A.B. Gas separation performance of 6FDA-DAM-ZIF-11 mixed-matrix membranes for H_2/CH_4 and CO_2/CH_4 separation. *Sep. Purif. Technol.* **2017**, *173*, 269–279. [CrossRef]

79. Hao, L.; Liao, K.-S.; Chung, T.-S. Photo-oxidative PIM-1 based mixed matrix membranes with superior gas separation performance. *J. Mater. Chem. A* **2015**, *3*, 17273–17281. [CrossRef]

80. Bae, T.; Lee, J.S.; Qiu, W.; Koros, W.J.; Jones, C.W.; Nair, S. A High-Performance Gas-Separation Membrane Containing Submicrometer-Sized Metal–Organic Framework Crystals. *Angew. Chemie Int. Ed.* **2010**, *49*, 9863–9866. [CrossRef] [PubMed]

81. Zhang, Q.; Luo, S.; Weidman, J.R.; Guo, R. Preparation and gas separation performance of mixed-matrix membranes based on triptycene-containing polyimide and zeolite imidazole framework (ZIF-90). *Polymer (Guildf)* **2017**, *131*, 209–216. [CrossRef]

82. DeStefano, M.R.; Islamoglu, T.; Garibay, S.J.; Hupp, J.T.; Farha, O.K. Room-Temperature Synthesis of UiO-66 and Thermal Modulation of Densities of Defect Sites. *Chem. Mater.* **2017**, *29*, 1357–1361. [CrossRef]

83. Cavka, J.H.; Jakobsen, S.; Olsbye, U.; Guillou, N.; Lamberti, C.; Bordiga, S.; Lillerud, K.P. A new zirconium inorganic building brick forming metal organic frameworks with exceptional stability. *J. Am. Chem. Soc.* **2008**, *130*, 13850–13851. [CrossRef] [PubMed]

84. Friebe, S.; Geppert, B.; Steinbach, F.; Caro, J. Metal–Organic Framework UiO-66 Layer: A Highly Oriented Membrane with Good Selectivity and Hydrogen Permeance. *ACS Appl. Mater. Interfaces* **2017**, *9*, 12878–12885. [CrossRef] [PubMed]

85. Kolokolov, D.I.; Stepanov, A.G.; Guillerm, V.; Serre, C.; Frick, B.; Jobic, H. Probing the Dynamics of the Porous Zr Terephthalate UiO-66 Framework Using 2H NMR and Neutron Scattering. *J. Phys. Chem. C* **2012**, *116*, 12131–12136. [CrossRef]

86. Shen, J.; Liu, G.; Huang, K.; Li, Q.; Guan, K.; Li, Y.; Jin, W. UiO-66-polyether block amide mixed matrix membranes for CO_2 separation. *J. Memb. Sci.* **2016**, *513*, 155–165. [CrossRef]

87. Anjum, M.W.; Vermoortele, F.; Khan, A.L.; Bueken, B.; De Vos, D.E.; Vankelecom, I.F.J. Modulated UiO-66-based mixed-matrix membranes for CO_2 separation. *ACS Appl. Mater. Interfaces* **2015**, *7*, 25193–25201. [CrossRef] [PubMed]

88. Venna, S.R.; Lartey, M.; Li, T.; Spore, A.; Kumar, S.; Nulwala, H.B.; Luebke, D.R.; Rosi, N.L.; Albenze, E. Fabrication of MMMs with improved gas separation properties using externally-functionalized MOF particles. *J. Mater. Chem. A* **2015**, *3*, 5014–5022. [CrossRef]

89. Khdhayyer, M.R.; Esposito, E.; Fuoco, A.; Monteleone, M.; Giorno, L.; Jansen, J.C.; Attfield, M.P.; Budd, P.M. Mixed matrix membranes based on UiO-66 MOFs in the polymer of intrinsic microporosity PIM-1. *Sep. Purif. Technol.* **2017**, *173*, 304–313. [CrossRef]

90. Ghalei, B.; Sakurai, K.; Kinoshita, Y.; Wakimoto, K.; Isfahani, A.P.; Song, Q.; Doitomi, K.; Furukawa, S.; Hirao, H.; Kusuda, H.; et al. Enhanced selectivity in mixed matrix membranes for CO_2 capture through efficient dispersion of amine-functionalized MOF nanoparticles. *Nat. Energy* **2017**, *2*, 17086. [CrossRef]

91. Zamidi Ahmad, M.; Navarro, M.; Lhotka, M.; Zornoza, B.; Téllez, C.; Fila, V.; Coronas, J. Enhancement of CO_2/CH_4 separation performances of 6FDA-based co-polyimides mixed matrix membranes embedded with UiO-66 nanoparticles. *Sep. Purif. Technol.* **2018**, *192*, 465–474. [CrossRef]

92. Ahmad, M.Z.; Navarro, M.; Lhotka, M.; Zornoza, B.; Téllez, C.; de Vos, W.M.; Benes, N.E.; Konnertz, N.M.; Visser, T.; Semino, R.; et al. Enhanced gas separation performance of 6FDA-DAM based mixed matrix membranes by incorporating MOF UiO-66 and its derivatives. *J. Memb. Sci.* **2018**, *558*, 64–77. [CrossRef]

93. Yazaydın, A.O.; Benin, A.I.; Faheem, S.A.; Jakubczak, P.; Low, J.J.; Willis, R.R.; Snurr, R.Q. Enhanced CO_2 adsorption in metal-organic frameworks via occupation of open-metal sites by coordinated water molecules. *Chem. Mater.* **2009**, *21*, 1425–1430. [CrossRef]

94. Du, M.; Li, L.; Li, M.; Si, R. Adsorption mechanism on metal organic frameworks of Cu-BTC, Fe-BTC and ZIF-8 for CO_2 capture investigated by X-ray absorption fine structure. *RSC Adv.* **2016**, *6*, 62705–62716. [CrossRef]

95. Ge, L.; Zhou, W.; Rudolph, V.; Zhu, Z. Mixed matrix membranes incorporated with size-reduced Cu-BTC for improved gas separation. *J. Mater. Chem. A* **2013**, *1*, 6350–6358. [CrossRef]

96. Abedini, R.; Mosayebi, A.; Mokhtari, M. Improved CO_2 separation of azide cross-linked PMP mixed matrix membrane embedded by nano-CuBTC metal organic framework. *Process Saf. Environ. Prot.* **2018**, *114*, 229–239. [CrossRef]

97. Tayebeh, K.; Mohammadreza, O.; Serge, K.; Denis, R. Amine-functionalized CuBTC/poly(ether-b-amide-6) (Pebax® MH 1657) mixed matrix membranes for CO_2/CH_4 separation. *Can. J. Chem. Eng.* **2017**, *95*, 2024–2033. [CrossRef]

98. Perez, E.V; Balkus, K.J.; Ferraris, J.P.; Musselman, I.H. Metal-organic polyhedra 18 mixed-matrix membranes for gas separation. *J. Memb. Sci.* **2014**, *463*, 82–93. [CrossRef]

99. Zhang, Y.; Musselman, I.H.; Ferraris, J.P.; Balkus, K.J. Gas permeability properties of Matrimid® membranes containing the metal-organic framework Cu–BPY–HFS. *J. Memb. Sci.* **2008**, *313*, 170–181. [CrossRef]

100. Sandrine, B.; Philip, L.L.; Christian, S.; Franck, M.; Thierry, L.; Gérard, F. Different Adsorption Behaviors of Methane and Carbon Dioxide in the Isotypic Nanoporous Metal Terephthalates MIL-53 and MIL-47. *J. Am. Chem. Soc.* **2005**, *127*, 13519–13521. [CrossRef]

101. Hu, Y.H.; Zhang, L. Hydrogen Storage in Metal-Organic Frameworks. *Adv. Mater.* **2010**, *22*, E117–E130. [CrossRef] [PubMed]

102. Adhikari, A.K.; Lin, K.-S.; Tu, M.-T. Hydrogen storage capacity enhancement of MIL-53(Cr) by Pd loaded activated carbon doping. *J. Taiwan Inst. Chem. Eng.* **2016**, *63*, 463–472. [CrossRef]

103. Lin, K.-S.; Adhikari, A.K.; Tu, M.-T.; Wang, C.-H.; Chiang, C.-L. Preparation, characterization, and hydrogen storage capacity of MIL-53 metal-organic frameworks. *J. Nanosci. Nanotechnol.* **2013**, *13*, 2549–2556. [CrossRef] [PubMed]

104. Mulder, F.M.; Assfour, B.; Huot, J.; Dingemans, T.J.; Wagemaker, M.; Ramirez-Cuesta, A.J. Hydrogen in the Metal−Organic Framework Cr MIL-53. *J. Phys. Chem. C* **2010**, *114*, 10648–10655. [CrossRef]

105. Hamon, L.; Llewellyn, P.L.; Devic, T.; Ghoufi, A.; Clet, G.; Guillerm, V.; Pirngruber, G.D.; Maurin, G.; Serre, C.; Driver, G.; et al. Co-adsorption and Separation of CO_2-CH_4 Mixtures in the Highly Flexible MIL-53(Cr) MOF. *J. Am. Chem. Soc.* **2009**, *131*, 17490–17499. [CrossRef] [PubMed]

106. Llewellyn, P.L.; Bourrelly, S.; Serre, C.; Filinchuk, Y.; Férey, G. How Hydration Drastically Improves Adsorption Selectivity for CO_2 over CH_4 in the Flexible Chromium Terephthalate MIL-53. *Angew. Chemie Int. Ed.* **2006**, *45*, 7751–7754. [CrossRef] [PubMed]

107. Hamon, L.; Serre, C.; Devic, T.; Loiseau, T.; Millange, F.; Férey, G.; Weireld, G. De Comparative Study of Hydrogen Sulfide Adsorption in the MIL-53(Al, Cr, Fe), MIL-47(V), MIL-100(Cr), and MIL-101(Cr) Metal−Organic Frameworks at Room Temperature. *J. Am. Chem. Soc.* **2009**, *131*, 8775–8777. [CrossRef] [PubMed]

108. Dorosti, F.; Omidkhah, M.; Abedini, R. Fabrication and characterization of Matrimid/MIL-53 mixed matrix membrane for CO_2/CH_4 separation. *Chem. Eng. Res. Des.* **2014**, *92*, 2439–2448. [CrossRef]

109. Hsieh, J.O.; Balkus, K.J.; Ferraris, J.P.; Musselman, I.H. MIL-53 frameworks in mixed-matrix membranes. *Microporous Mesoporous Mater.* **2014**, *196*, 165–174. [CrossRef]

110. Abedini, R.; Omidkhah, M.; Dorosti, F. Highly permeable poly(4-methyl-1-pentyne)/NH_2-MIL 53 (Al) mixed matrix membrane for CO_2/CH_4 separation. *RSC Adv.* **2014**, *4*, 36522–36537. [CrossRef]

111. Feijani, E.A.; Mahdavi, H.; Tavasoli, A. Poly(vinylidene fluoride) based mixed matrix membranes comprising metal organic frameworks for gas separation applications. *Chem. Eng. Res. Des.* **2015**, *96*, 87–102. [CrossRef]

112. Ahmadi Feijani, E.; Tavasoli, A.; Mahdavi, H. Improving Gas Separation Performance of Poly(vinylidene fluoride) Based Mixed Matrix Membranes Containing Metal–Organic Frameworks by Chemical Modification. *Ind. Eng. Chem. Res.* **2015**, *54*, 12124–12134. [CrossRef]

113. Tien-Binh, N.; Vinh-Thang, H.; Chen, X.Y.; Rodrigue, D.; Kaliaguine, S. Polymer functionalization to enhance interface quality of mixed matrix membranes for high CO_2/CH_4 gas separation. *J. Mater. Chem. A* **2015**, *3*, 15202–15213. [CrossRef]

114. Zhu, H.; Jie, X.; Wang, L.; Liu, D.; Cao, Y. Polydimethylsiloxane/postmodified MIL-53 composite layer coated on asymmetric hollow fiber membrane for improving gas separation performance. *J. Appl. Polym. Sci.* **2017**, *134*. [CrossRef]

115. Hong, D.-Y.; Hwang, Y.K.; Serre, C.; Férey, G.; Chang, J.-S. Porous Chromium Terephthalate MIL-101 with Coordinatively Unsaturated Sites: Surface Functionalization, Encapsulation, Sorption and Catalysis. *Adv. Funct. Mater.* **2009**, *19*, 1537–1552. [CrossRef]

116. Naseri, M.; Mousavi, S.F.; Mohammadi, T.; Bakhtiari, O. Synthesis and gas transport performance of MIL-101/Matrimid mixed matrix membranes. *J. Ind. Eng. Chem.* **2015**, *29*, 249–256. [CrossRef]

117. Rajati, H.; Navarchian, A.H.; Tangestaninejad, S. Preparation and characterization of mixed matrix membranes based on Matrimid/PVDF blend and MIL-101(Cr) as filler for CO_2/CH_4 separation. *Chem. Eng. Sci.* **2018**, *185*, 92–104. [CrossRef]

118. Zhang, W.; Liu, D.; Guo, X.; Huang, H.; Zhong, C. Fabrication of mixed-matrix membranes with MOF-derived porous carbon for CO_2 separation. *AIChE J.* **2018**. [CrossRef]

119. Tanh Jeazet, H.B.; Sorribas, S.; Román-Marín, J.M.; Zornoza, B.; Téllez, C.; Coronas, J.; Janiak, C. Increased Selectivity in CO_2/CH_4 Separation with Mixed-Matrix Membranes of Polysulfone and Mixed-MOFs MIL-101(Cr) and ZIF-8. *Eur. J. Inorg. Chem.* **2016**, *2016*, 4363–4367. [CrossRef]

120. Xie, K.; Fu, Q.; Webley, P.A.; Qiao, G.G. MOF Scaffold for a High-Performance Mixed-Matrix Membrane. *Angew. Chemie Int. Ed.* **2018**. [CrossRef]

121. Shahid, S.; Nijmeijer, K. High pressure gas separation performance of mixed-matrix polymer membranes containing mesoporous Fe(BTC). *J. Memb. Sci.* **2014**, *459*, 33–44. [CrossRef]

122. Rita, A.; Ribeiro, R.P.P.L.; Mota, J.P.B.; Alves, V.D.; Esteves, I.A.A.C. Separation and Purification Technology CO_2/N_2 gas separation using Fe(BTC)-based mixed matrix membranes: A view on the adsorptive and filler properties of metal-organic frameworks. *Sep. Purif. Technol.* **2018**, *202*, 174–184. [CrossRef]

123. Dorosti, F.; Alizadehdakhel, A. Fabrication and investigation of PEBAX/Fe-BTC, a high permeable and CO_2 selective mixed matrix membrane. *Chem. Eng. Res. Des.* **2017**, *136*, 119–128. [CrossRef]

124. Cadiau, A.; Adil, K.; Bhatt, P.M.; Belmabkhout, Y.; Eddaoudi, M. A metal-organic framework–based splitter for separating propylene from propane. *Science* **2016**, *353*, 137–140. [CrossRef] [PubMed]

125. Chen, K.; Xu, K.; Xiang, L.; Dong, X.; Han, Y.; Wang, C.; Sun, L.; Pan, Y. Enhanced CO_2/CH_4 separation performance of mixed-matrix membranes through dispersion of sorption-selective MOF nanocrystals. *J. Membr. Sci.* **2018**, *563*, 360–370. [CrossRef]

126. Bae, T.-H.; Long, J.R. CO_2/N_2 separations with mixed-matrix membranes containing Mg_2(dobdc) nanocrystals. *Energy Environ. Sci.* **2013**, *6*, 3565–3569. [CrossRef]

127. Smith, Z.P.; Bachman, J.E.; Li, T.; Gludovatz, B.; Kusuma, V.A.; Xu, T.; Hopkinson, D.P.; Ritchie, R.O.; Long, J.R. Increasing M_2(dobdc) Loading in Selective Mixed-Matrix Membranes: A Rubber Toughening Approach. *Chem. Mater.* **2018**, *30*, 1484–1495. [CrossRef]

128. Konstas, K.; Taylor, J.W.; Thornton, A.W.; Doherty, C.M.; Lim, W.X.; Bastow, T.J.; Kennedy, D.F.; Wood, C.D.; Cox, B.J.; Hill, J.M. Lithiated porous aromatic frameworks with exceptional gas storage capacity. *Angew. Chemie Int. Ed.* **2012**, *51*, 6639–6642. [CrossRef] [PubMed]

129. Dechnik, J.; Gascon, J.; Doonan, C.; Janiak, C.; Sumby, C.J. Mixed-matrix membranes. *Angew. Chemie Int. Ed.* **2017**, *56*, 9292–9310. [CrossRef] [PubMed]

130. Thomas, J.M.H.; Trewin, A. Amorphous PAF-1: Guiding the rational design of ultraporous materials. *J. Phys. Chem. C* **2014**, *118*, 19712–19722. [CrossRef]

131. Ben, T.; Ren, H.; Ma, S.; Cao, D.; Lan, J.; Jing, X.; Wang, W.; Xu, J.; Deng, F.; Simmons, J.M. Targeted synthesis of a porous aromatic framework with high stability and exceptionally high surface area. *Angew. Chem.* **2009**, *121*, 9621–9624. [CrossRef]

132. Li, Y.; Roy, S.; Ben, T.; Xu, S.; Qiu, S. Micropore engineering of carbonized porous aromatic framework (PAF-1) for supercapacitors application. *Phys. Chem. Chem. Phys.* **2014**, *16*, 12909–12917. [CrossRef] [PubMed]

133. Fu, J.; Wu, J.; Custelcean, R.; Jiang, D. Nitrogen-doped porous aromatic frameworks for enhanced CO_2 adsorption. *J. Colloid Interface Sci.* **2015**, *438*, 191–195. [CrossRef] [PubMed]

134. Ben, T.; Qiu, S. Porous aromatic frameworks: Synthesis structure and functions. *CrystEngComm* **2013**, *15*, 17–26. [CrossRef]

135. Lau, C.H.; Nguyen, P.T.; Hill, M.R.; Thornton, A.W.; Konstas, K.; Doherty, C.M.; Mulder, R.J.; Bourgeois, L.; Liu, A.C.Y.; Sprouster, D.J. Ending aging in super glassy polymer membranes. *Angew. Chem. Int. Ed.* **2014**, *53*, 5322–5326. [CrossRef] [PubMed]

136. Volkov, A.V; Bakhtin, D.S.; Kulikov, L.A.; Terenina, M.V.; Golubev, G.S.; Bondarenko, G.N.; Legkov, S.A.; Shandryuk, G.A.; Volkov, V.V; Khotimskiy, V.S. Stabilization of gas transport properties of PTMSP with porous aromatic framework: Effect of annealing. *J. Memb. Sci.* **2016**, *517*, 80–90. [CrossRef]

137. Lau, C.H.; Konstas, K.; Doherty, C.M.; Kanehashi, S.; Ozcelik, B.; Kentish, S.E.; Hill, A.J.; Hill, M.R. Tailoring Physical Aging in Super Glassy Polymers with Functionalized Porous Aromatic Frameworks for CO_2 Capture Tailoring Physical Aging in Super Glassy Polymers with Functionalized Porous Aromatic Frameworks for CO_2 Capture. *Chem. Mater.* **2015**, *27*, 4756–4762. [CrossRef]

138. Mitra, T.; Bhavsar, R.S.; Adams, D.J.; Budd, P.M.; Cooper, A.I. PIM-1 mixed matrix membranes for gas separations using cost-effective hypercrosslinked nanoparticle fillers. *Chem. Commun.* **2016**, *52*, 5581–5584. [CrossRef] [PubMed]

139. Yaumi, A.L.; Bakar, M.Z.A.; Hameed, B.H. Recent advances in functionalized composite solid materials for carbon dioxide capture. *Energy* **2017**, *124*, 461–480. [CrossRef]

140. Matsukata, M.; Sawamura, K.; Sekine, Y.; Kikuchi, E. Chapter 8—Review on Prospects for Energy Saving in Distillation Process with Microporous Membranes. In *Inorganic Polymeric and Composite Membranes*; Oyama, S.T., Stagg-Williams, S., Eds.; Elsevier: New York, NY, USA, 2011; Volume 14, pp. 175–193, ISBN 0927-5193.

141. Rangnekar, N.; Mittal, N.; Elyassi, B.; Caro, J.; Tsapatsis, M. Zeolite membranes—A review and comparison with MOFs. *Chem. Soc. Rev.* **2015**, *44*, 7128–7154. [CrossRef] [PubMed]

142. Bastani, D.; Esmaeili, N.; Asadollahi, M. Polymeric mixed matrix membranes containing zeolites as a filler for gas separation applications: A review. *J. Ind. Eng. Chem.* **2013**, *19*, 375–393. [CrossRef]

143. Hosseinzadeh Beiragh, H.; Omidkhah, M.; Abedini, R.; Khosravi, T.; Pakseresht, S. Synthesis and characterization of poly (ether-block-amide) mixed matrix membranes incorporated by nanoporous ZSM-5 particles for CO_2/CH_4 separation. *Asia-Pac. J. Chem. Eng.* **2016**, *11*, 522–532. [CrossRef]

144. Dorosti, F.; Omidkhah, M.; Abedini, R. Enhanced CO_2/CH_4 separation properties of asymmetric mixed matrix membrane by incorporating nano-porous ZSM-5 and MIL-53 particles into Matrimid® 5218. *J. Nat. Gas Sci. Eng.* **2015**, *25*, 88–102. [CrossRef]

145. Bryan, N.; Lasseuguette, E.; Van Dalen, M.; Permogorov, N.; Amieiro, A.; Brandani, S.; Ferrari, M.C. Development of mixed matrix mebranes containing zeolites for post-combustion carbon capture. *Energy Procedia* **2014**, *63*, 160–166. [CrossRef]

146. Bezerra, D.P.; da Silva, F.W.M.; de Moura, P.A.S.; Sousa, A.G.S.; Vieira, R.S.; Rodriguez-Castellon, E.; Azevedo, D.C.S. CO_2 adsorption in amine-grafted zeolite 13X. *Appl. Surf. Sci.* **2014**, *314*, 314–321. [CrossRef]

147. Surya Murali, R.; Ismail, A.F.; Rahman, M.A.; Sridhar, S. Mixed matrix membranes of Pebax-1657 loaded with 4A zeolite for gaseous separations. *Sep. Purif. Technol.* **2014**, *129*, 1–8. [CrossRef]

148. Zhao, D.; Ren, J.; Li, H.; Hua, K.; Deng, M. Poly(amide-6-b-ethylene oxide)/SAPO-34 mixed matrix membrane for CO_2 separation. *J. Energy Chem.* **2014**, *23*, 227–234. [CrossRef]

149. Hong, M.; Li, S.; Falconer, J.L.; Noble, R.D. Hydrogen purification using a SAPO-34 membrane. *J. Memb. Sci.* **2008**, *307*, 277–283. [CrossRef]

150. Rezakazemi, M.; Shahidi, K.; Mohammadi, T. Hydrogen separation and purification using crosslinkable PDMS/zeolite A nanoparticles mixed matrix membranes. *Int. J. Hydrog. Energy* **2012**, *37*, 14576–14589. [CrossRef]

151. Atalay-Oral, Ç.; Tokay, B.; Erdem-enatalar, A.; Tantekin-ŞErsolmaz, B. Ferrierite-poly(vinyl acetate) mixed matrix membranes for gas separation: A comparative study. *Microporous Mesoporous Mater.* **2018**, *259*, 17–25. [CrossRef]

152. Fernández-Barquín, A.; Casado-Coterillo, C.; Palomino, M.; Valencia, S.; Irabien, A. LTA/Poly(1-trimethylsilyl-1-propyne) Mixed-Matrix Membranes for High-Temperature CO_2/N_2 Separation. *Chem. Eng. Technol.* **2015**, *38*, 658–666. [CrossRef]

153. Fernández-Barquín, A.; Casado-Coterillo, C.; Palomino, M.; Valencia, S.; Irabien, A. Permselectivity improvement in membranes for CO_2/N_2 separation. *Sep. Purif. Technol.* **2016**, *157*, 102–111. [CrossRef]

154. Gong, H.; Lee, S.S.; Bae, T.-H. Mixed-matrix membranes containing inorganically surface-modified 5A zeolite for enhanced CO_2/CH_4 separation. *Microporous Mesoporous Mater.* **2017**, *237*, 82–89. [CrossRef]

155. Pakizeh, M.; Hokmabadi, S. Experimental study of the effect of zeolite 4A treated with magnesium hydroxide on the characteristics and gas-permeation properties of polysulfone-based mixed-matrix membranes. *J. Appl. Polym. Sci.* **2017**, *134*. [CrossRef]

156. Sanaeepur, H.; Nasernejad, B. Advances Aminosilane-functionalization of a nanoporous Y-type zeolite for application in a cellulose acetate based mixed matrix membrane for CO_2 separation. *RSC Adv.* **2014**, *4*, 63966–63976. [CrossRef]

157. Peng, Y.; Li, Y.; Ban, Y.; Jin, H.; Jiao, W.; Liu, X.; Yang, W. Metal-organic framework nanosheets as building blocks for molecular sieving membranes. *Science* **2014**, *346*, 1356–1359. [CrossRef] [PubMed]

158. Zhong, Z.; Yao, J.; Chen, R.; Low, Z.; He, M.; Liu, J.Z.; Wang, H. Oriented two-dimensional zeolitic imidazolate framework-L membranes and their gas permeation properties. *J. Mater. Chem. A* **2015**, *3*, 15715–15722. [CrossRef]

159. Roth, W.J.; Nachtigall, P.; Morris, R.E.; Čejka, J. Two-Dimensional Zeolites: Current Status and Perspectives. *Chem. Rev.* **2014**, *114*, 4807–4837. [CrossRef] [PubMed]

160. Jeon, M.Y.; Kim, D.; Kumar, P.; Lee, P.S.; Rangnekar, N.; Bai, P.; Shete, M.; Elyassi, B.; Lee, H.S.; Narasimharao, K.; et al. Ultra-selective high-flux membranes from directly synthesized zeolite nanosheets. *Nature* **2017**, *543*, 690. [CrossRef] [PubMed]

161. Varoon, K.; Zhang, X.; Elyassi, B.; Brewer, D.D.; Gettel, M.; Kumar, S.; Lee, J.A.; Maheshwari, S.; Mittal, A.; Sung, C.-Y.; et al. Dispersible Exfoliated Zeolite Nanosheets and Their Application as a Selective Membrane. *Science* **2011**, *334*, 72–75. [CrossRef] [PubMed]

162. Liu, G.; Jiang, Z.; Cao, K.; Nair, S.; Cheng, X.; Zhao, J.; Gomaa, H.; Wu, H.; Pan, F. Pervaporation performance comparison of hybrid membranes filled with two-dimensional ZIF-L nanosheets and zero-dimensional ZIF-8 nanoparticles. *J. Memb. Sci.* **2017**, *523*, 185–196. [CrossRef]

163. Kim, W.; Lee, J.S.; Bucknall, D.G.; Koros, W.J.; Nair, S. Nanoporous layered silicate AMH-3/cellulose acetate nanocomposite membranes for gas separations. *J. Memb. Sci.* **2013**, *441*, 129–136. [CrossRef]

164. Kang, Z.; Peng, Y.; Qian, Y.; Yuan, D.; Addicoat, M.A.; Heine, T.; Hu, Z.; Tee, L.; Guo, Z.; Zhao, D. Mixed Matrix Membranes (MMMs) Comprising Exfoliated 2D Covalent Organic Frameworks (COFs) for Efficient CO_2 Separation. *Chem. Mater.* **2016**, *28*, 1277–1285. [CrossRef]

165. Rodenas, T.; Luz, I.; Prieto, G.; Seoane, B.; Miro, H.; Corma, A.; Kapteijn, F.; Llabrés i Xamena, F.X.; Gascon, J. Metal–organic framework nanosheets in polymer composite materials for gas separation. *Nat. Mater.* **2014**, *14*, 48. [CrossRef] [PubMed]

166. Shete, M.; Kumar, P.; Bachman, J.E.; Ma, X.; Smith, Z.P.; Xu, W.; Mkhoyan, K.A.; Long, J.R.; Tsapatsis, M. On the direct synthesis of Cu(BDC) MOF nanosheets and their performance in mixed matrix membranes. *J. Memb. Sci.* **2018**, *549*, 312–320. [CrossRef]

167. Cheng, Y.; Wang, X.; Jia, C.; Wang, Y.; Zhai, L.; Wang, Q.; Zhao, D. Ultrathin mixed matrix membranes containing two-dimensional metal-organic framework nanosheets for efficient CO_2/CH_4 separation. *J. Memb. Sci.* **2017**, *539*, 213–223. [CrossRef]

168. Yang, Y.; Goh, K.; Wang, R.; Bae, T.-H. High-performance nanocomposite membranes realized by efficient molecular sieving with CuBDC nanosheets. *Chem. Commun.* **2017**, *53*, 4254–4257. [CrossRef] [PubMed]

169. Kang, Z.; Peng, Y.; Hu, Z.; Qian, Y.; Chi, C.; Yeo, L.Y.; Tee, L.; Zhao, D. Mixed matrix membranes composed of two-dimensional metal-organic framework nanosheets for pre-combustion CO_2 capture: A relationship study of filler morphology versus membrane performance. *J. Mater. Chem. A* **2015**, *3*, 20801–20810. [CrossRef]

membranes
MDPI

Article

Estimating CO₂/N₂ Permselectivity through Si/Al = 5 Small-Pore Zeolites/PTMSP Mixed Matrix Membranes: Influence of Temperature and Topology

Clara Casado-Coterillo [1,*], Ana Fernández-Barquín [1], Susana Valencia [2] and Ángel Irabien [1]

[1] Department of Chemical and Biomolecular Engineering, Universidad de Cantabria, Av. Los Castros s/n, 39005 Santander, Spain; fbarquina@unican.es (A.F.-B.); irabienj@unican.es (Á.I.)
[2] Instituto de Tecnología Química, Universitat Politècnica de València-Consejo Superior de Investigaciones Científicas, Av. de los Naranjos s/n, 46022 Valencia, Spain; svalenci@itq.upv.es
* Correspondence: casadoc@unican.es; Tel.: +34-942-206-777

Received: 11 May 2018; Accepted: 15 June 2018; Published: 16 June 2018

Abstract: In the present work, the effect of zeolite type and topology on CO_2 and N_2 permeability using zeolites of different topology (CHA, RHO, and LTA) in the same Si/Al = 5, embedded in poly(trimethylsilyl-1-propyne) (PTMSP) is evaluated with temperature. Several models are compared on the prediction of CO_2/N_2 separation performance and then the modified Maxwell models are selected. The CO_2 and N_2 permeabilities through these membranes are predicted with an average absolute relative error (AARE) lower than 0.6% taking into account the temperature and zeolite loading and topology on non-idealities such as membrane rigidification, zeolite–polymer compatibility and sieve pore blockage. The evolution of this structure–performance relationship with temperature has also been predicted.

Keywords: mixed matrix membranes; Poly(trimethylsilyl-1-propyne) (PTMSP); small-pore zeolites (CHA, RHO, LTA); temperature; modeling

1. Introduction

Carbon capture strategies are still envisaged as one of the major challenges for preventing CO_2 emissions to the atmosphere from anthropogenic sources. Membrane separation technology is often presented as an energy efficient and economical alternative to conventional capture technologies although not yet passing through the stage of pilot plant scale [1]. Polymer membranes for CO_2 separation are especially constrained by a performance 'upper bound' trade-off between gas permeability and selectivity, which becomes especially significant for treating large volumes of flue gas. The simultaneous improvement on membrane permeability and selectivity is very attractive for industrial applications. Mixed matrix membranes (MMMs), which consist of the introduction of small amounts, usually below 30 wt %, of a special filler providing properties such as a molecular sieve, ion-exchange and robustness in a processable polymer matrix [2], are surpassing this upper bound [3–7]. More than homogenous distribution, the main challenge of MMM fabrication is achieving a good adhesion and compatibility between the inorganic filler and the polymer, avoiding the voids and defects that deteriorate separation performance [8].

Polyimide materials have been, firstly, studied for gas separation because of their stability and selectivity. However, permeability is usually low for CO_2 separation [9]. The first and most widely used fillers are zeolites since the pioneering work of Zimmermann et al. [10]. Recently, zeolite 5A was introduced in Matrimid to prepare MMMs for CO_2/CH_4 separation, after particle surface modification to obtain a defect-free membrane [11]. Amooghin et al. [12] reported the ion exchange effect of Ag^+ in zeolite Y-filled

Matrimid MMMs led to a CO_2 permeability increase of 123% from 8.64 Barrer in pure Matrimid to 18 Barrer in 15% AgY-filled MMM, where 1 Barrer is defined as 10^{-10} cm^3(STP) cm cm^{-2} s^{-1} $cmHg^{-1}$.

A simple approach to produce high permeability and selectivity membranes without the use of modifiers that complicate the synthesis procedures is the variation of the inorganic particles composition themselves to influence the polarity in comparison with the selected polymer matrix. In the case of zeolites, this is represented by the Si/Al ratio and determines many properties of the material, including ion exchange capacity [13]. Thus, for the development of high perm-selective membrane materials for CO_2 separation, we focused on the most permeable polymer, poly(trimethylsilyl-1-propyne), PTMSP, and observed that the adhesion with LTA fillers and therefore CO_2/N_2 separation properties were best with a low Si/Al ratio even upon increasing temperature [14]. The strong influence of zeolite topology on CO_2 adsorption has also been acknowledged [15], giving the possibility to locally tune the energy interactions, promoting size and shape selectivity and clustering. However, this effect is not always straightforward because most zeolites cannot be synthesized in pure silica form or at similar Si/Al compositions. Exceptions to this rule are LTA (ITQ-29) [16] and CHA [17]. To avoid this and to see that the lower Si/Al favored the compatibility with glassy hydrophobic PTMSP [14], we fixed an intermediate value of the Si/Al ratio to 5, in order to study the influence of the zeolite filler topology using different small pore zeolites (LTA, CHA, RHO) in the CO_2/N_2 separation of PTMSP-based MMMs in the temperature range 298–333 K [18]. These MMM surpassed the Robeson's upper bound at 5 wt % loading even at increasing temperature, but the separation of CO_2/N_2 mixtures with a 12.5 wt % CO_2 content resulted in a real separation factor much lower than the intrinsic selectivity of the membrane material.

Besides the large number of research and publications devoted to new MMM material combinations for gas separation, there is also a growing literature on the development of systematic approaches to describe gas transport through MMMs [19–21]. The MMM performance has been evaluated as a function of the membrane morphology imposed by the filler loading and several models have been compared lately [22–25]. They all present several limitations such as not being valid but at low filler loadings, a large number of adjustable parameters, or not being able to predict the non-idealities common in MMM morphologies that influence their gas separation performance. The most accurate models reported so far are those proposed by Moore et al. [26] and Li et al. [27], accounting for the void interphase, which describes the compatibility between the zeolite filler and the polymer continuous matrix, and the polymer chain rigidification caused by the effect of the inorganic particles embedded in the polymer matrix, in the first case. The second one distinguishes the transport of fast and slow gas molecules, respectively, and introduces the effect of pore blockage that may become important when the dispersed phase is a porous particle as zeolites are [25]. In fact, partial pore blockage has been recently proven to be the dominant effect when porous zeolites are used as fillers in Matrimid, impeding the increase of permeability with increasing dispersed phase loading [28], in agreement with most studies dealing with low permeability polyimides like Matrimid, polysulfone (PSf), and polyethersulfone (PES). The effect of temperature in the performance of those models is seldom reported [29,30].

Thus, in this work the gas permeation through MMMs prepared from small pore zeolites of different topology and constant Si/Al = 5 in PTMSP is evaluated by modified Maxwell models including the void thickness, chain immobilization and pore-blockage effects, and their variation with temperature.

2. Materials and Methods

The MMMs were prepared by a solution-casting method from PTMSP (ABCR, Gelest) previously dissolved in toluene, and CHA, RHO and LTA zeolites of Si/Al = 5 prepared at the Instituto de Tecnología Química (UPV-CSIC) as reported in our previous work [18]. The characteristics of the zeolite fillers used in this work are summarized in Table 1. The membranes were stored in plastic Petri dishes and they were immersed in methanol for a few minutes before gas permeation experiments to remove the effect of aging [31]. The density of the PTMSP pure membranes is 0.75 g/cm^3.

Table 1. Properties of the zeolite fillers with Si/Al = 5 used in this work.

Filler	Crystal Size (μm)	Density (g/cm^3)	Pore Size [1] (nm)	Structure [2]
LTA	0.5	1.498 [32]	0.41	
CHA	1.0	2.090	0.38	
RHO	1.5	1.442 [33]	0.36	

[1] From [18]. [2] The crystallographic structures have been taken from the International Zeolite Database (http://www.iza-structure.org/databases/): View of the planes 100 for LTA and 001 for CHA and RHO, respectively.

Figure 1 shows the high magnification scanning electron microscope (SEM) images of 5 wt % CHA, LTA, and RHO/PTMSP MMMs. As reported in a previous work [18], the smaller LTA particles are dispersed throughout the whole membrane thickness, of which a small glimpse can be seen in Figure 1a, while the larger CHA and RHO zeolites form a bottom layer of particles bound together by the polymer, as observed in Figure 1b for a CHA/PTMSP MMM. In the case of RHO, this adhesion is so strong that individual crystals are not easily discerned in Figure 1c. In this work, we want to focus on the compatibility and adhesion between the filler and the polymer, as the main challenge in MMM fabrication [34,35], thus it is important to notice in Figure 1 that even the largest particles at the bottom of the membrane are apparently well adhered with the polymer continuous matrix.

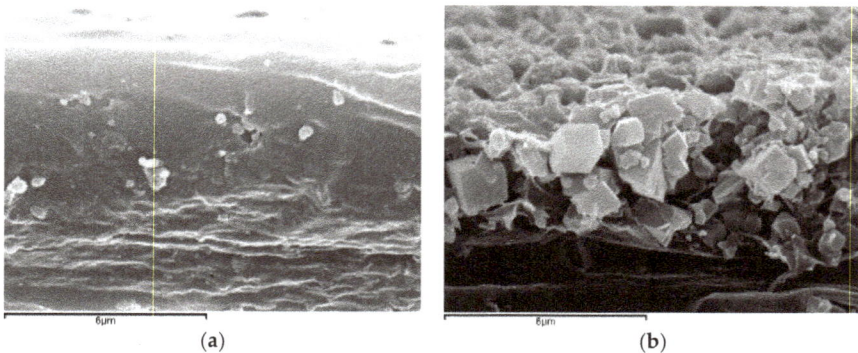

(a)

(b)

Figure 1. *Cont.*

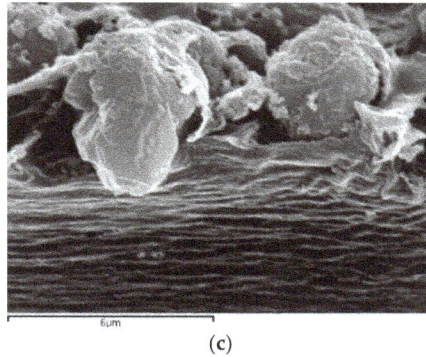

(c)

Figure 1. Scanning electron microscope (SEM) images of the detailed contact between LTA (**a**); CHA (**b**); RHO (**c**) and poly(trimethylsilyl-1-propyne) (PTMSP) in 5 wt % loaded mixed matrix membranes (MMMs). Bars correspond to 6 μm.

The thickness of every MMM is measured experimentally at 5 points over the membrane surface for each membrane sample using a IP-65 Mitutoyo digital micrometer (Kawasaki, Japan) with a precision of 0.001 mm. The average thickness for all the MMMs tested in this work was 75 ± 14 μm.

The single gas permeation of N_2 and CO_2 was measured in that order, using a home-made constant volume set-up described elsewhere [14,18], in the temperature range 298 to 333 K and a feed pressure of 3–4 bar and atmospheric permeate pressure. The average values of the permeabilities and selectivities obtained previously and used in this work are collected in Table A1 in Appendix A.

3. Results and Discussion

3.1. Comparison of Known Mixed-Matrix Membrane Model Predictions

First, well-known models for predicting MMM permeation (Appendix B) have been compared in terms of the percentage average absolute relative error (AARE) with the permeability of CO_2 and N_2 through MMMs, as

$$\mathrm{AARE}(\%) = \frac{100}{N} \sum_{i=1}^{N} \left| \frac{P_i^{calc} - P_i^{exp}}{P_i^{exp}} \right| \tag{1}$$

where N is the number of experimental data points [23].

A Maxwell model often represents the ideal case with no defects and no distortion of separation properties. Table 2 summarizes the AARE values obtained with the models most commonly encountered in the literature, averaged for the whole range of temperature studied in our laboratory to allow comparison.

Table 2. Percentage of average absolute relative error (AARE) for CO_2 and N_2 permeation (first and second values in every entry) prediction, highlighting those AARE values lower than 20%.

MMM	Series	Parallel	Maxwell	Higuchi	Felske	Lewis-Nielsen
5CHA/PTMSP	**17.32**/370	108/2026	106/2006	146/2609	118/32.4	24.9/**2.14**
10CHA/PTMSP	24.2/143	102/2966	99.7/2909	96.8/2854	80/936	$10^{-4}/10^{-5}$
5LTA/PTMSP	20.6/33.3	**11.8**/516	**11.4**/498	26.3/708	**2.54/10^{-3}**	**0.46/0.01**
10LTA/PTMSP	40.9/50.0	**14.5**/631	**4.79**/214	**14.6**/560	67.4/**9.04**	**3.98/10^{-5}**
20LTA/PTMSP	45.0/50.0	**7.11**/212	**8.28**/198	**10.4**/194	**3.00/10^{-4}**	**4.37/10^{-5}**
5RHO/PTMSP	**8.62**/126	**12.7**/362	**12.4**/357	**16.7**/395	**0.85/6·10^{-4}**	**1.84/0.6·10^{-5}**
10RHO/PTMSP	24.0/216	57.0/1030	54.5/1003	49.3/947	**0.03/2·10^{-3}**	**4.32/0.02**
20RHO/PTMSP	45.3/52.4	72.2/947	63.8/892	44.2/756	**22.0/5·10^{-4}**	**12.3/10^{-4}**

According to Table 2, N_2 permeability values cannot be predicted by the series, parallel, Maxwell and Higuchi models with acceptable error in all the range of temperature under study. The prediction accuracy of CO_2 permeability varies as a function of the zeolite topology. Regarding CO_2 permeability, the series and parallel model approaches fit the 5 wt % CHA/PTMSP MMM performance at 323 K, with a lower average AARE for this membrane. The CO_2 permeability of LTA/PTMSP MMMs can be described by parallel, Maxwell and Higuchi models in the whole range of operating temperatures and LTA loadings, while the series model only fits the experimental data at low loading. As for the RHO/PTMSP MMM, this is only valid up to 10 wt % RHO loading in the PTMSP matrix. This agrees with the data reported for other MMMs prepared with dispersed fillers of RHO topology [36] where the Maxwell equation only describes the CO_2 permeability at low loading, as observed for the ZIF-20/Matrimid MMM, being ZIF-20 a zeolite imidazolate framework of RHO topology as well [36]. In the case of our RHO/PTMSP MMMs, all previous models overestimate the experimental permeabilities.

Only the model predictions with AARE lower than 20% are represented in Figure 2, for clarification purposes. The original Maxwell equation overestimates the experimental value for the permeability of all gases and membranes, especially for N_2 permeability. This overestimation is more significant at lower operation temperatures, as reported by Clarizia et al. [14]. In this work, this is true for CHA/PTMSP MMMs with the series model, Figure 2a, and the parallel and Maxwell model for LTA/PTMSP MMMs, Figure 2c. These are simplifications of the general Maxwell equation expressed by Equation (B1) to predict the overall steady-state permeability through an ideal defect-free MMM [26]. Those models provide a simple, quantitative framework to predict the transport properties of MMM when the transport properties of the constituent phases are known, especially at low dispersed phase loading. Only more advanced modifications of this Maxwell equation, such as Felske and Lewis–Nielsen, provide enough accuracy for the description of MMM performance, especially in the case of the slow permeating gas, N_2, as reflected in Figure 2b,d,f.

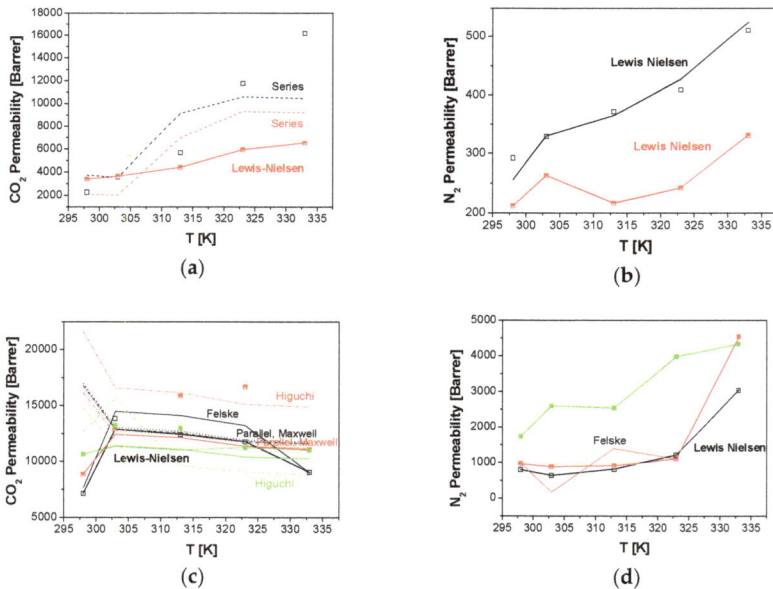

Figure 2. *Cont.*

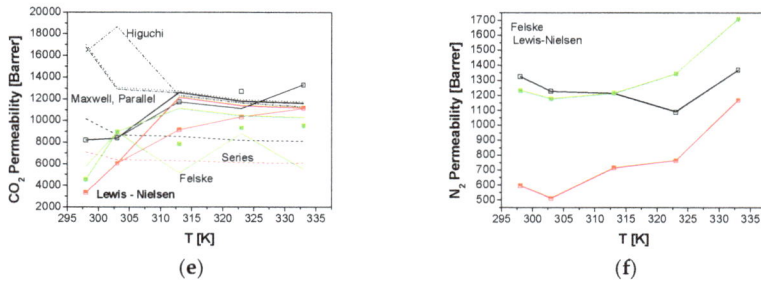

Figure 2. Comparison of CO_2 (left) and N_2 permeabilities through CHA (**a,b**), LTA (**c,d**) and RHO (**e,f**)/PTMSP MMMs with the predictions by the series (dashed lines), parallel (dotted lines), original Maxwell (dash-dot), Higuchi (dash dot dot), Felske (thin continuous line) and Lewis–Nielsen (thick continuous line) models, as a function of temperature. Zeolite loading: 5 wt % (black), 10 wt % (red), 20 wt % (green).

3.2. Reduced Mobility Modified Maxwell Model

In order to account for the non-idealities in the membrane morphology accounting for the compatibility that influence the membrane performance [30], polymer chain rigidification and interphase void thickness, the Maxwell model is applied twice to predict the permeability of a pseudo-interphase induced by the interfacial contact between filler and polymer matrix [25], as schematized in Figure 3a.

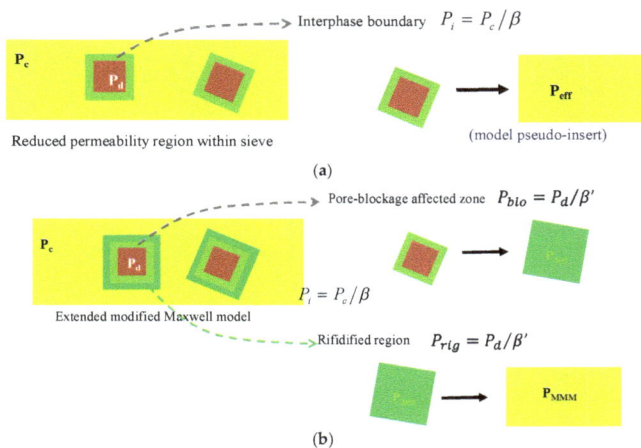

Figure 3. Schemes of the modified Maxwell model proposed by Moore et al. [26] (**a**) and the extended modified Maxwell model proposed by Li et al. [27] (**b**), both adapted for this work.

According to the reduced mobility modified Maxwell model, the effective permeability through the pseudo-insert in Figure 3a, P_{eff}, is calculated first by

$$P_{eff} = P_I \left[\frac{P_d + 2P_I - 2\varphi_s(P_c - P_d)}{P_d + 2P_I + \varphi_s(P_c - P_d)} \right] \qquad (2)$$

where φ_d is the filler volume fraction in the polymer matrix, P_I is the permeability through the rigidified continuous matrix, calculated as the ratio between the experimental permeability through a pure PTMSP membrane [18] and an adjustable parameter, β, as described in Figure 3a, and P_d is the permeability through the zeolite. In this work, this value has been taken from literature data on CO_2 and N_2 permeation through pure zeolite membranes of similar Si/Al ratio and topology (Table 3) to avoid the usual dispersion on this parameter when calculated from experimental solubility isotherms [23].

Table 3. Permeability data of the pure zeolite dispersed phase, P_d, used for the model predictions.

Zeolite Dispersed Phase	$P_d(CO_2)$ (Barrer)	$P_d(N_2)$ (Barrer)	T (K)	Reference
CHA (Si/Al = 5) [1]	88	0.59	293	[37]
CHA (pure silica)	539	55	313	[38]
LTA (Si/Al = 1)	139	0.048	298	[25]
RHO [2]	623	260	298	[33]

[1] Si/Al = 5 as the zeolites used in this work. [2] The CO_2 permeabilities reported for ZIF-8 composite values are considered as the Rho here, given the similar sodalite topology.

In Equation (2), P_I acts as the permeability of the continuous phase, considering as such the interphase, assuming the bulk of the zeolite as the dispersed phase and the affected zeolite interphase with reduced permeability as the continuous phase [39], as represented in the scheme in Figure 3a. φ_s is the volume fraction of the dispersed sieve phase in combined sieve and interphase, given by

$$\varphi_s = \frac{\varphi_d}{\varphi_d + \varphi_I} = \frac{r_d^3}{(r_d + l_I)^3} \tag{3}$$

where φ_I is the volume fraction of the interface, and l_I is the thickness of the 'interface void'. The permeability of the whole MMM is thus estimated by applying the Maxwell equation again, as

$$P_{MMM} = P_c \left[\frac{P_{eff} + 2P_c - 2\varphi_s(P_c - P_{eff})}{P_{eff} + 2P_c + \varphi_s(P_c - P_{eff})} \right] \tag{4}$$

As $\varphi_d + \varphi_I$ increases to one, the interphases of neighboring dispersed particles overlap and the overall mixed matrix is rigidified. This occurs preferentially as the zeolite particle loading is increased or the interphase void distance is increased, i.e., voids appear because embedding in the polymer chains becomes more difficult.

Equations (2)–(4) predict the overall performance of MMMs taking into account the case morphologies identified by Moore et al. [26], adapted to distinguish the performance of the fast and slow gas in CO_2/N_2 separation, and including the influence of temperature. This model is thus based on three adjustable parameters, the interphase thickness, l_I, and the chain immobilization factor, β, which depends on the permeating gas molecule [39], whose values are presented in Tables 4–6 for the CHA/PTMSP, LTA/PTMSP and RHO/PTMSP MMM, respectively.

Table 4. Parameters estimated by the reduced mobility modified Maxwell model for the CHA/PTMSP MMMs.

T (K)	5 wt %		10 wt %	
	l_I (μm) = 1.39		l_I (μm) = 0.98	
	β (CO_2)	β (N_2)	β (CO_2)	β (N_2)
298	7.42	61.2	4.90	86.61
303	4.56	53.28	3.48	64.0
313	2.25	42.8	2.87	70.5
323	1.01	31.41	1.97	50.4
333	0.73	20.5	1.00	10.2

Table 5. Parameters estimated by the reduced mobility modified Maxwell model for the LTA/PTMSP MMMs.

T (K)	5 wt % l_I (μm) = 0.60		10 wt % l_I (μm) = 0.56 ± 0.08		20 wt % l_I (μm) = 0.27	
	β (CO₂)	β (N₂)	β (CO₂)	β (N₂)	β (CO₂)	β (N₂)
298	2.35	21.9	1.83	17.4	1.39	8.82
303	0.93	27.1	1.00	12.0	0.86	5.84
313	1.01	18.9	0.80	11.0	0.85	5.37
323	1.00	10.2	0.72	8.34	0.92	2.72
333	1.29	3.38	1.06	2.49	0.93	2.08

Table 6. Parameters estimated by the reduced mobility modified Maxwell model for the RHO/PTMSP MMMs.

T (K)	5 wt % l_I (μm) = 1.76		10 wt % l_I (μm) = 1.23		20 wt % l_I (μm) = 0.79	
	β (CO₂)	β (N₂)	β (CO₂)	β (N₂)	β (CO₂)	β (N₂)
298	2.06	0.31	10.62	1.95	3.36	1.46
303	1.57	0.35	2.10	2.98	1.28	1.54
313	1.07	0.30	1.33	1.29	1.43	1.33
323	0.91	0.28	1.17	0.93	1.12	0.93
333	0.87	0.17	1.01	0.45	1.08	0.58

As expected, the chain immobilization factor, β, is smaller for CO_2 than N_2. This confirms that the polymer chain rigidification normally results in a larger resistance to the transport of the gas with larger molecular diameter [27]. The RHO/PTMSP MMM revealed a different trend, although only at 298 K, which may be attributed to the agglomeration of these larger crystal size and smaller pore size particles at the bottom of the MMM. Interestingly, $\beta(CO_2)$ and $\beta(N_2)$ of the three types of MMMs converge to similar values upon increasing temperature. This may be attributed to the compensating effects of polymer flexibility and chain rigidification of the polymer matrix, which are accentuated for the larger size of the RHO particles than LTA and CHA. This agrees with the current statement that in gas separation through MMMs there is not only an optimum in zeolite loading but also in operating temperature [40].

The thickness of the interphase between the zeolite and the polymer matrix, l_I (μm), accounts for the compatibility between the zeolite and polymer phases, as well as the defects or voids due to poor compatibility between zeolites and polymer [25]. In this work, the void thickness decreases with increasing zeolite loading and is independent of the type of gas and temperature. It can also be observed that this parameter l_I is influenced by the zeolite topology, in the following order: l_I (LTA/PTMSP) < l_I (CHA/PTMSP) < l_I (RHO/PTMSP). This is attributed to the different interaction with the polymer matrix, and the decreasing particle size, in agreement with results obtained for zeolite-APTES/PES MMMs [27]. Those authors obtained as thickness of the rigidified region l_I = 0.30 μm for a cubic zeolite A (Si/Al = 1) dispersed phase in PES, and values of the chain immobilization factor (β) of 3 and 4, for O_2 and N_2, respectively. A rigidified thickness of 1.4 μm and chain immobilization factor was reported for ZIF-20/polysulfone MMMs, estimating a P_d = 45 Barrer, in agreement with pore ZIF membranes of similar pore size and topology [41]. Therefore, the magnitude of the adjustable parameters obtained in this work are in the same order of magnitude.

These parameters allow a prediction of the permeability through these MMMs by this model with an error of up to a global AARE below 6 ± 1%, where the maximum errors lie on 10CHA/PTMSP and 10RHO/PTMSP membranes at 298 K.

3.3. Extended Pore-Blockage Reduced Mobility Modified Maxwell Model

Although in this work the channel opening of the zeolites (0.38, 0.41 and 0.36 nm for CHA, LTA and RHO topologies, respectively) lie in the same range as the gas pair molecules to be separated, we have included the analysis of the partial pore blockage effect [25,35] as Li et al. [27] for zeolite A-APTES/PES MMM, adapted in the Scheme shown in Figure 3b. This approach consists in applying the Maxwell equation not just twice, but three times, and requires not just three, but six adjustable parameters, in order to define the dispersed phase volume fraction in the pore-blockage and the rigidified region, as well as the immobilization factor for the pair of gases in both sections.

Firstly, the permeability in the pore-blockage affected zone near the zeolite particle surface as represented in Figure 3b, is calculated by

$$P_{3rd} = P_{blo}\left[\frac{P_d + 2(P_d/\beta') - 2\varphi_3((P_d/\beta') - P_d)}{P_d + 2(P_d/\beta') + \varphi_3((P_d/\beta') - P_d)}\right] \tag{5}$$

Secondly, the P_{3rd} permeability calculated by Equation (5) is entered as the new dispersed phase, and the permeability of the rigidified region, P_{rig}, is taken as the continuous phase, to calculate the new P_{eff}, P_{2nd}:

$$P_{2nd} = P_{rig}\left[\frac{P_{3rd} + 2(P_c/\beta) - 2\varphi_2((P_c/\beta) - P_{3rd})}{P_{3rd} + 2(P_c/\beta) + \varphi_2((P_c/\beta) - P_{3rd})}\right] \tag{6}$$

Thirdly and lastly, the permeability through the bulk of the MMM is calculated using P_{2nd} as the new permeability for the dispersed phase, turning the previous equations into

$$P_{MMM} = P_c\left[\frac{P_{2nd} + 2P_c - 2(\varphi_d + \varphi_{blo} + \varphi_{rig})(P_c - P_{2nd})}{P_{2nd} + 2P_c + (\varphi_d + \varphi_{blo} + \varphi_{rig})(P_c - P_{2nd})}\right] \tag{7}$$

with

$$\varphi_3 = \frac{\varphi_d}{\varphi_d + \varphi_{blo}} \tag{8}$$

and

$$\varphi_2 = \frac{\varphi_d + \varphi_{blo}}{\varphi_d + \varphi_{blo} + \varphi_{rig}} \tag{9}$$

Now, the adjustable parameters are φ_{blo} and φ_{rig}, the calculated volume fraction of the pore-blockage affected region, and the rigidified region, respectively, as well as β' and β, whose values depend on the permeating gas, and identify the partial pore blockage affected and rigidified polymer region, respectively, as given in Figure 3b. Note that β is similar to the chain immobilization factor introduced by the previous reduced mobility modified Maxwell model, discussed in the previous section.

Figures 4–6 show the comparison of the prediction of CO_2 and N_2 permeability using both modified Maxwell models. The experimental results are well described for the Si/Al = 5 zeolites, indicating a good compatibility between intermediate Si/Al zeolites and the glassy PTMSP [14]. The optimized β value is higher for N_2 than CO_2, for CHA and RHO/PTMSP MMMs. $\beta(N_2)$ values of 0.92 are obtained for the CHA/PTMSP MMMs, independently of zeolite loading, where as they increase from 0.66 to 1.40 for the RHO/PTMSP MMMs. $\beta(CO_2)$ gives smaller values than $\beta(N_2)$, as expected for smaller molecules. $\beta(CO_2)$ follows similar trends as $\beta(N_2)$, being constant for CHA and LTA/PTMSP MMMs, at values of 0.3 and 0.2, respectively, and increasing from 0.26 to 0.94 with increasing loading for RHO/PTMSP MMMs. These values are smaller than 1.6, the value recently published for Sigma-1/Matrimid MMMs, considering also the partial pore blockage effect [28]. The values of $\beta'(CO_2)$ are 0.06 for CHA and RHO/PTMSP MMMs, and below 0.03 for LTA/PTMSP MMMs. The $\beta'(N_2)$ are 70% higher in the LTA and RHO/PTMSP MMMs, and 30% higher than $\beta'(CO_2)$ in the case of CHA/PTMSP MMMs. These results reveal that, although the partial pore blockage is low in small–pore zeolites, it is more significant for the smaller pore size zeolite fillers as CHA or RHO, than LTA.

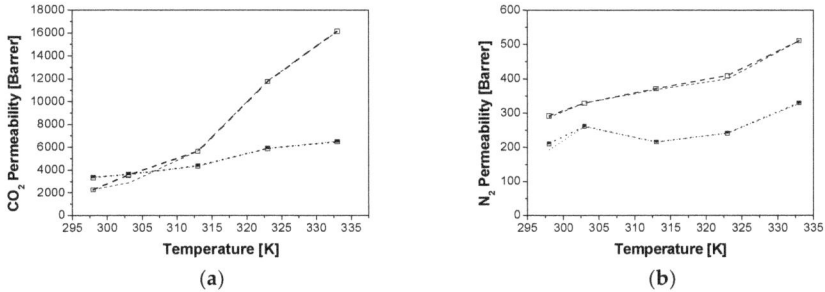

Figure 4. Effect of temperature and zeolite loading on the CO_2 (**a**) and N_2 (**b**) permeability through CHA/PTMSP MMMs: Thin lines correspond to the reduced mobility modified Maxwell model and thick lines to the extended modified Maxwell model. Dash, dot and continuous patterns, and void, half-filled and full symbols, refer to 5 wt %, 10 wt % and 20 wt % zeolite loading, respectively.

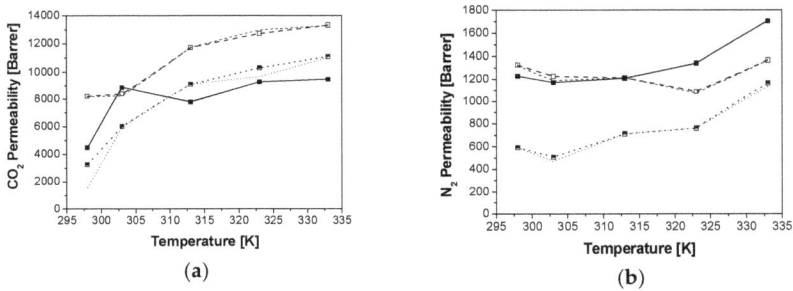

Figure 5. Effect of temperature and zeolite loading on the CO_2 (**a**) and N_2 (**b**) permeability through LTA/PTMSP MMMs: Thin lines correspond to the reduced mobility modified model and thick lines to the extended modified Maxwell model. Dash, dot and continuous patterns, and void, half-filled and full symbols, refer to 5 wt %, 10 wt % and 20 wt % zeolite loading, respectively.

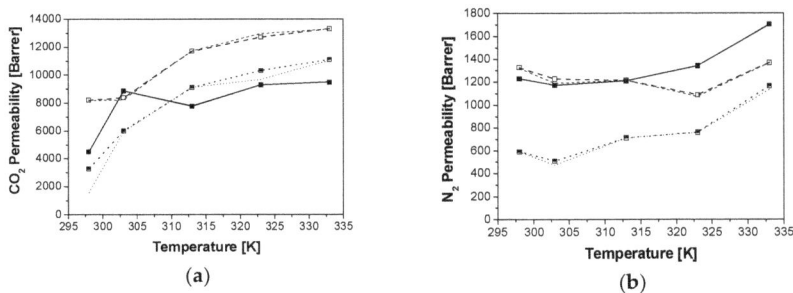

Figure 6. Effect of temperature and zeolite loading on the CO_2 (**a**) and N_2 (**b**) permeability through RHO/PTMSP MMMs: Thin lines correspond to the reduced mobility modified model and thick lines to the extended modified Maxwell model. Dash, dot and continuous patterns, and void, half-filled and full symbols, refer to 5 wt %, 10 wt % and 20 wt % zeolite loading, respectively.

The models describe well the CO_2 and N_2 permeability through the Si/Al = 5 zeolite/PTMSP MMMs as a function of zeolite loading, topology and temperature. The CO_2 permeability increases with temperature while the N_2 permeability slightly increases for CHA and RHO/PTMSP MMMs, behavior similar to pure zeolite membranes, as reflected by the activation energies derived from the Arrhenius equation in the previous work [18], in agreement with other works in literature [42]. The LTA/PTMSP MMMs show a maximum performance at 10 wt % zeolite loading and 323 K, losing permselectivity at higher loading and temperature. The worst AARE for the prediction of experimental permeabilities through the extended partial pore blockage reduced mobility model is 0.6%, for the 5 wt % CHA/MMM at 313 K, which were in some of the best agreement with the first modified Maxwell model. Partial pore blockage may be affecting permeability even with small-pore zeolite fillers in a glassy polymer matrix [28].

4. Conclusions

The experimental CO_2 and N_2 permeabilities of Si/Al = 5 small-pore zeolites/PTMSP MMM has been compared with modified Maxwell model predictions as a function of zeolite topology (CHA, LTA, RHO), loading (0–20 wt %) and temperature (298–333 K). Three adjustable parameters accounting for the membrane rigidification, void interphase and partial pore-blockage have been optimized at values lower than reported in literature. They reveal the compatibility between Si/Al = 5 zeolites dispersed in the glassy polymer PTMSP, as well as a small influence of partial pore blockage in the case of the smaller pore size CHA and RHO. The CO_2 and N_2 permeabilities through these membranes are predicted with an AARE lower than 0.6% taking into account zeolite loading and topology on non-idealities such as membrane rigidification and sieve pore blockage and their influence on MMM performance. The evolution of this structure-performance relationship with temperature has also been predicted. The implementation of the Arrhenius dependency of the MMM permeability and the prediction studied in this work constitute a step further towards the understanding of the MMM performance in order to develop new membrane materials and module configurations with potential application in CO_2 separation, which will be addressed in a future work.

Author Contributions: Conceptualization, C.C.-C.; Data curation, A.F.-B.; Funding acquisition, A.I.; Investigation, C.C.-C., A.F.-B., S.V. and A.I.; Methodology, C.C.-C.; Supervision, A.I.; Writing—original draft, C.C.-C.; Writing—review and editing, C.C.-C., S.V. and A.I.

Funding: This research was funded by Spanish MINECO—General Secretariat for Research, Development and Innovation under project CTQ2016-76231-C2-1-R at the University of Cantabria, and MAT2015-71842-P, at the Instituto de Tecnología Química.

Acknowledgments: The authors gratefully acknowledge the financial support of the Spanish MINECO—General Secretariat for Research, Development and Innovation under project CTQ2016-76231-C2-1-R at the University of Cantabria, and MAT2015-71842-P, at the Instituto de Tecnología Química. Miguel Palomino is thanked for the acquisition of the SEM images at the Electron Microscopy Service of the Universitat Politècnica de València.

Conflicts of Interest: The authors declare no conflict of interest.

Appendix A

The experimental permeation data obtained in a previous work [18] are collected in Table A1.

Table A1. Experimental data of the different MMMs with increasing order of particle size (LTA, 0.5 μm; CHA, 1 μm; RHO, 1.5 μm).

Filler and Loading [18]	T (K)	P(CO_2) (Barrer)	P(N_2) (Barrer)	$\alpha(CO_2/N_2)$
	298	7150	794	9
	303	13,881	637	22
5 wt % LTA	313	12,448	816	15
	323	11,770	1208	10
	333	9026	3044	3

Table A1. *Cont.*

Filler and Loading [18]	T (K)	P(CO$_2$) (Barrer)	P(N$_2$) (Barrer)	α(CO$_2$/N$_2$)
10wt % LTA	298	8813	951	9
	303	12,921	865	15
	313	15,802	892	18
	323	16,648	1078	15
	333	11,029	4520	2.5
20 wt % LTA	298	10,587	1720	6
	303	13,178	2585	5
	313	12,980	2519	5
	323	11,175	3966	3
	333	10,964	4316	2.5
5 wt % CHA	298	2274	292	8
	303	3575	329	11
	313	5651	372	15
	323	11,772	409	29
	333	16,145	511	32
10 wt % CHA	298	3363	211	16
	303	3620	262	14
	313	4351	216	20
	323	5892	241	24
	333	6485	330	20
5 wt % RHO	298	8205	1325	6
	303	8383	1227	7
	313	11,722	1214	10
	323	12,726	1089	12
	333	13,324	1368	10
10 wt % RHO	298	3262	592	6
	303	5996	509	12
	313	9111	712	13
	323	10,304	761	14
	333	11,114	1166	10
20 wt % RHO	298	4479	1229	4
	303	8883	1173	8
	313	7784	1210	6
	323	9293	1341	7
	333	9498	1704	6

Appendix B

The MMM performance has been evaluated as a function of the membrane morphology imposed by the filler loading using several models that have been compared lately [20,23–25]. Equation (A1) was derived by Maxwell for semi-conductors and is widely accepted as an easy tool for a quick estimation of the performance of MMMs from phase-separated blends [3,30]:

$$P_{\mathrm{mmm}} = P_{\mathrm{c}} \left[\frac{P_{\mathrm{d}} + 2P_{\mathrm{c}} - 2\varphi_{\mathrm{d}}(P_{\mathrm{c}} - P_{\mathrm{d}})}{P_{\mathrm{d}} + 2P_{\mathrm{c}} + \varphi_{\mathrm{d}}(P_{\mathrm{c}} - P_{\mathrm{d}})} \right] \tag{A1}$$

where φ_{d} is the dispersed phase volume fraction, calculated from the nominal weight fraction of the zeolite in the MMMs, using the density of the PTMSP polymer and the corresponding zeolite density (Table 1).

The minimum value of effective permeability of a given penetrant in a MMM is given by considering a series mechanism of transport through the dispersed and continuous phases (Equation (A2)):

$$P_{mmm} = \frac{P_c P_d}{(1 - \varphi_d)P_d + \varphi_d P_c} \tag{A2}$$

and the maximum value is taken when both phases are assumed to contribute in parallel to the flow direction (Equation (A3)):

$$P_{mmm} = \varphi_d P_d + (1 - \varphi_d)P_c \tag{A3}$$

Other important models used for the description of gas permeation in MMMs are the Higuchi, Felske and Lewis–Nielsen, Bruggemann and Pal models [20]. The last two are not presented in this work because they are implicit equations derived from Maxwell and Lewis–Nielsen that have to be solved numerically.

The Higuchi model is applied for a random dispersion of spherical filler particles but lacks mathematical rigor [24]. The main equation for porous zeolite particle fillers is given by:

$$P_{mmm} = P_c \left[1 + \frac{3\varphi_d}{\frac{P_d + 2P_c}{P_d - P_c} - \varphi_d - K\left[\frac{(1 - \varphi_d)(P_d - P_c)}{P_d + 2P_c} \right]} \right] \tag{B4}$$

where K is an empirical constant containing shape description, with no physical meaning. In this work, it only adjusts the accepted value of 0.78 for 5 wt % CHA, 5–10 wt % LTA5/PTMSP. 10 wt % CHA/PTMSP is adjusted to $K = 0.999$ and for the rest of the membranes K varies randomly between 0.0001 and 0.03 at different temperatures.

The Felske model was originally used for the description of the thermal conductivity of composites of core-shell particles (core particle covered with interfacial layer) and also for permeability measurement. It gives almost the same predictions as the modified Maxwell model and it can be reduced to Maxwell's when the interfacial layer is absent [25]. It is described by Equations (A5)–(A7), as

$$P_{mmm} = P_c \left[\frac{2(1 - \varphi_d) + (1 + 2\varphi_d)(\beta/\gamma)}{(2 + \varphi_d) + (1 - \varphi_d)(\beta/\gamma)} \right] \tag{A5}$$

with

$$\beta = \frac{(2 + \delta^3)P_d - 2(1 - \delta^3)P_I}{P_c} = (2 + \delta^3)\frac{P_d}{P_c} - 2(1 - \delta^3)\frac{P_I}{P_c} \tag{A6}$$

and

$$\gamma = 1 + 2\delta^3 - (1 - \delta^3)\frac{P_d}{P_c} \tag{A7}$$

where $\delta = r_I/r_d$. This model also needs three adjustable parameters, as in the reduced mobility modified Maxwell model.

The Lewis–Nielsen model was originally proposed for describing an elastic modulus of particulate composites, and the following equation can be used to predict the effective permeability in MMMs:

$$P_{mmm} = P_c \left[\frac{1 + 2\varphi_d(\alpha - 1)/(\alpha + 2)}{1 - \psi\varphi_d(\alpha - 1)/(\alpha + 2)} \right] \tag{A8}$$

where

$$\psi = 1 + \left(\frac{1 - \varphi_m}{\varphi_m} \right) \tag{A9}$$

This model might represent a correct definition of the permeability over the range of $0 < \varphi_d < \varphi_m$. The solution diverges when $\varphi_d = \varphi_m$ and it should be noted that when $\varphi_m \to 1$, the Lewis–Nielsen model reduces to the Maxwell equation (Equation (A1)).

Membranes **2018**, *8*, 32

References

1. Bhown, A.S. Status and analysis of next generation post-combustion CO_2 capture technologies. *Energy Procedia* **2014**, *63*, 542–549. [CrossRef]
2. Dong, G.; Li, H.; Chen, V. Challenges and opportunities for mixed-matrix membranes for gas separation. *J. Mater. Chem. A* **2013**, *1*, 4610. [CrossRef]
3. Robeson, L.M. Polymer blends in membrane transport processes. *Ind. Eng. Chem. Res.* **2010**, *49*, 11859–11865. [CrossRef]
4. Jusoh, N.; Fong Yeong, Y.; Leng Chew, T.; Keong Lau, K.; Mohd Shariff, A. Current development and challenges of mixed matrix membranes for CO_2/CH_4 separation. *Sep. Purif. Rev.* **2016**, *454*, 321–344. [CrossRef]
5. Rezakazemi, M.; Ebadi Amooghin, A.; Montazer-Rahmati, M.M.; Ismail, A.F.; Matsuura, T. State-of-the-art membrane based CO_2 separation using mixed matrix membranes (MMMs): An overview on current status and future directions. *Prog. Polym. Sci.* **2014**, *39*, 817–861. [CrossRef]
6. Zhang, Y.; Sunarso, J.; Liu, S.; Wang, R. Current status and development of membranes for CO_2/CH_4 separation: A review. *Int. J. Greenh. Gas Control* **2013**, *12*, 84–107. [CrossRef]
7. Rahaman, M.S.A.; Cheng, L.H.; Xu, X.H.; Zhang, L.; Chen, H.L. A review of carbon dioxide capture and utilization by membrane integrated microalgal cultivation processes. *Renew. Sustain. Energy Rev.* **2011**, *15*, 4002–4012. [CrossRef]
8. Bastani, D.; Esmaeili, N.; Asadollahi, M. Polymeric mixed matrix membranes containing zeolites as a filler for gas separation applications: A review. *J. Ind. Eng. Chem.* **2013**, *19*, 375–393. [CrossRef]
9. Powell, C.E.; Qiao, G.G. Polymeric CO_2/N_2 gas separation membranes for the capture of carbon dioxide from power plant flue gases. *J. Membr. Sci.* **2006**, *279*, 1–49. [CrossRef]
10. Zimmerman, C.M.; Singh, A.; Koros, W.J. Tailoring mixed matrix composite membranes for gas separations. *J. Membr. Sci.* **1997**, *137*, 145–154. [CrossRef]
11. Gong, H.; Lee, S.S.; Bae, T.H. Mixed-matrix membranes containing inorganically surface-modified 5A zeolite for enhanced CO_2/CH_4 separation. *Microporous Mesoporous Mater.* **2017**, *237*, 82–89. [CrossRef]
12. Ebadi Amooghin, A.; Omidkhah, M.; Sanaeepur, H.; Kargari, A. Preparation and characterization of Ag+ ion-exchanged zeolite-Matrimid??5218 mixed matrix membrane for CO_2/CH_4 separation. *J. Energy Chem.* **2016**, *25*, 450–462. [CrossRef]
13. Lopes, A.C.; Martins, P.; Lanceros-Mendez, S. Aluminosilicate and aluminosilicate based polymer composites: Present status, applications and future trends. *Prog. Surf. Sci.* **2014**, *89*, 239–277. [CrossRef]
14. Fernández-Barquín, A.; Casado-Coterillo, C.; Palomino, M.; Valencia, S.; Irabien, A. LTA/Poly(1-trimethylsilyl-1-propyne) Mixed-Matrix Membranes for High-Temperature CO_2/N_2 Separation. *Chem. Eng. Technol.* **2015**, *38*, 658–666. [CrossRef]
15. Pera-titus, M. Porous Inorganic membranes for CO_2 capture: Present and prospects. *Chem. Rev.* **2014**, *114*, 1413–1492. [CrossRef] [PubMed]
16. Palomino, M.; Corma, A.; Rey, F.; Valencia, S. New insights on CO_2-methane separation using LTA zeolites with different Si/Al ratios and a first comparison with MOFs. *Langmuir* **2010**, *26*, 1910–1917. [CrossRef] [PubMed]
17. Hedin, N.; DeMartin, G.J.; Roth, W.J.; Strohmaier, K.G.; Reyes, S.C. PFG NMR self-diffusion of small hydrocarbons in high silica DDR, CHA and LTA structures. *Microporous Mesoporous Mater.* **2008**, *109*, 327–334. [CrossRef]
18. Fernández-Barquín, A.; Casado-Coterillo, C.; Palomino, M.; Valencia, S.; Irabien, A. Permselectivity improvement in membranes for CO_2/N_2 separation. *Sep. Purif. Technol.* **2016**, *157*, 102–111. [CrossRef]
19. Hashemifard, S.A.; Ismail, A.F.; Matsuura, T. A new theoretical gas permeability model using resistance modeling for mixed matrix membrane systems. *J. Membr. Sci.* **2010**, *350*, 259–268. [CrossRef]
20. Shimekit, B.; Mukhtar, H.; Murugesan, T. Prediction of the relative permeability of gases in mixed matrix membranes. *J. Membr. Sci.* **2011**, *373*, 152–159. [CrossRef]
21. Ebneyamini, A.; Azimi, H.; Tezel, F.H.; Thibault, J. Mixed matrix membranes applications: Development of a resistance-based model. *J. Membr. Sci.* **2017**, *543*, 351–360. [CrossRef]
22. Pal, R. Permeation models for mixed matrix membranes. *J. Colloid Interface Sci.* **2008**, *317*, 191–198. [CrossRef] [PubMed]

23. Hashemifard, S.A.; Ismail, A.F.; Matsuura, T. Prediction of gas permeability in mixed matrix membranes using theoretical models. *J. Membr. Sci.* **2010**, *347*, 53–61. [CrossRef]

24. Vinh-Thang, H.; Kaliaguine, S. Predictive models for mixed-matrix membrane performance: A review. *Chem. Rev.* **2013**, *113*, 4980–5028. [CrossRef] [PubMed]

25. Shen, Y.; Lua, A.C. Theoretical and experimental studies on the gas transport properties of mixed matrix membranes based on polyvinylidene fluoride. *AIChE J.* **2013**, 3–194. [CrossRef]

26. Moore, T.T.; Mahajan, R.; Vu, D.Q.; Koros, W.J. Hybrid membrane materials comprising organic polymers with rigid dispersed phases. *AIChE J.* **2004**, *50*, 311–321. [CrossRef]

27. Li, Y.; Guan, H.M.; Chung, T.S.; Kulprathipanja, S. Effects of novel silane modification of zeolite surface on polymer chain rigidification and partial pore blockage in polyethersulfone (PES)-zeolite A mixed matrix membranes. *J. Membr. Sci.* **2006**, *275*, 17–28. [CrossRef]

28. Gheimasi, K.M.; Peydayesh, M.; Mohammadi, T.; Bakhtiari, O. Prediction of CO_2/CH_4 permeability through Sigma-1-Matrimid®5218 MMMs using the Maxwell model. *J. Membr. Sci.* **2014**, *466*, 265–273. [CrossRef]

29. Rezaei-DashtArzhandi, M.; Ismail, A.F.; Ghanbari, M.; Bakeri, G.; Hashemifard, S.A.; Matsuura, T.; Moslehyani, A. An investigation of temperature effects on the properties and CO_2 absorption performance of porous PVDF/montmorillonite mixed matrix membranes. *J. Nat. Gas Sci. Eng.* **2016**, *31*, 515–524. [CrossRef]

30. Clarizia, G.; Algieri, C.; Drioli, E. Filler-polymer combination: A route to modify gas transport properties of a polymeric membrane. *Polymer* **2004**, *45*, 5671–5681. [CrossRef]

31. Hill, A.J.; Pas, S.J.; Bastow, T.J.; Burgar, M.I.; Nagai, K.; Toy, L.G.; Freeman, B.D. Influence of methanol conditioning and physical aging on carbon spin-lattice relaxation times of poly(1-trimethylsilyl-1-propyne). *J. Membr. Sci.* **2004**, *243*, 37–44. [CrossRef]

32. García, E.J.; Pérez-Pellitero, J.; Pirngruber, G.D.; Jallut, C.; Palomino, M.; Rey, F.; Valencia, S. Tuning the adsorption properties of zeolites as adsorbents for CO_2 separation: Best compromise between the working capacity and selectivity. *Ind. Eng. Chem. Res.* **2014**, *53*, 9860–9874. [CrossRef]

33. Diestel, L.; Liu, X.L.; Li, Y.S.; Yang, W.S.; Caro, J. Comparative permeation studies on three supported membranes: Pure ZIF-8, pure polymethylphenylsiloxane, and mixed matrix membranes. *Microporous Mesoporous Mater.* **2014**, *189*, 210–215. [CrossRef]

34. Mahajan, R.; Burns, R.; Schaeffer, M.; Koros, W.J. Challenges in forming successful mixed matrix membranes with rigid polymeric materials. *J. Appl. Polym. Sci.* **2002**, *86*, 881–890. [CrossRef]

35. Chung, T.S.; Jiang, L.Y.; Li, Y.; Kulprathipanja, S. Mixed matrix membranes (MMMs) comprising organic polymers with dispersed inorganic fillers for gas separation. *Prog. Polym. Sci.* **2007**, *32*, 483–507. [CrossRef]

36. Safak Boroglu, M.; Ugur, M.; Boz, I. Enhanced gas transport properties of mixed matrix membranes consisting of Matrimid and RHO type ZIF-12 particles. *Chem. Eng. Res. Des.* **2017**, *123*, 201–213. [CrossRef]

37. Wu, T.; Diaz, M.C.; Zheng, Y.; Zhou, R.; Funke, H.H.; Falconer, J.L.; Noble, R.D. Influence of propane on CO_2/CH_4 and N_2/CH_4 separations in CHA zeolite membranes. *J. Membr. Sci.* **2015**, *473*, 201–209. [CrossRef]

38. Kida, K.; Maeta, Y.; Yogo, K. Pure silica CHA-type zeolite membranes for dry and humidified CO_2/CH_4 mixtures separation. *Sep. Purif. Technol.* **2018**, *197*, 116–121. [CrossRef]

39. Li, Y.; Chung, T.S.; Cao, C.; Kulprathipanja, S. The effects of polymer chain rigidification, zeolite pore size and pore blockage on polyethersulfone (PES)-zeolite A mixed matrix membranes. *J. Membr. Sci.* **2005**, *260*, 45–55. [CrossRef]

40. Karkhanechi, H.; Kazemian, H.; Nazockdast, H.; Mozdianfard, M.R.; Bidoki, S.M. Fabrication of homogenous polymer-zeolite nanocomposites as mixed-matrix membranes for gas separation. *Chem. Eng. Technol.* **2012**, *35*, 885–892. [CrossRef]

41. Bux, H.; Liang, F.; Li, Y.; Cravillon, J.; Wiebcke, M.; Caro, J. Zeolitic Imidazolate Framework molecular sieving membrane titania support. *J. Am. Chem. Soc. Commun.* **2009**, *131*, 16000–16001. [CrossRef] [PubMed]

42. Li, S.; Jiang, X.; Yang, Q.; Shao, L. Effects of amino functionalized polyhedral oligomeric silsesquioxanes on cross-linked poly(ethylene oxide) membranes for highly-efficient CO_2 separation. *Chem. Eng. Res. Des.* **2017**, *122*, 280–288. [CrossRef]

membranes

MDPI

Article

Hydrolytic Degradation and Mechanical Stability of Poly(ε-Caprolactone)/Reduced Graphene Oxide Membranes as Scaffolds for In Vitro Neural Tissue Regeneration

Sandra Sánchez-González, Nazely Diban * and Ane Urtiaga

Department of Chemical and Biomolecular Engineering, University of Cantabria, Avda. Los Castros s/n, 39005 Santander, Spain; sandra.sanchez@unican.es (S.S.-G.); urtiaga@unican.es (A.U.)
* Correspondence: dibann@unican.es; Tel.: +34-942-206778

Received: 9 February 2018; Accepted: 1 March 2018; Published: 5 March 2018

Abstract: The present work studies the functional behavior of novel poly(ε-caprolactone) (PCL) membranes functionalized with reduced graphene oxide (rGO) nanoplatelets under simulated in vitro culture conditions (phosphate buffer solution (PBS) at 37 °C) during 1 year, in order to elucidate their applicability as scaffolds for in vitro neural regeneration. The morphological, chemical, and DSC results demonstrated that high internal porosity of the membranes facilitated water permeation and procured an accelerated hydrolytic degradation throughout the bulk pathway. Therefore, similar molecular weight reduction, from 80 kDa to 33 kDa for the control PCL, and to 27 kDa for PCL/rGO membranes, at the end of the study, was observed. After 1 year of hydrolytic degradation, though monomers coming from the hydrolytic cleavage of PCL diffused towards the PBS medium, the pH was barely affected, and the rGO nanoplatelets mainly remained in the membranes which envisaged low cytotoxic effect. On the other hand, the presence of rGO nanomaterials accelerated the loss of mechanical stability of the membranes. However, it is envisioned that the gradual degradation of the PCL/rGO membranes could facilitate cells infiltration, interconnectivity, and tissue formation.

Keywords: hydrolytic bulk degradation mechanism; in vitro human neural models; neural tissue regeneration; poly (ε-caprolactone); reduced graphene oxide

1. Introduction

In the recent years, there has been an increased interest in the functionalization of biocompatible polymers traditionally used for biomedical applications and FDA approved, such as poly(ε-caprolactone) (PCL), in order to incorporate different chemical, mechanical, or electrical stimuli, and therefore converting these polymers from plain cell supports to tissue regenerative inductive materials [1,2]. Different strategies have been used to introduce functional cues into polymers: loading with protein and growth factors [3,4], polymer blending or copolymerization [5,6], and the formation of composites with different types of nanomaterials [7–9] are among the most popular approaches. Particularly, while under the shadow of certain controversy, graphene and graphene derivatives, as graphene oxide (GO), have been studied to exploit their peculiar properties: capacity to interact with biomolecules, cells, and tissues, and to enhance the mechanical, electrical, and/or magnetic properties of polymer–graphene composite materials [10,11]. Moreover, graphene has demonstrated the potential to direct differentiation into neural cell lineages of numerous stem cell types, such as embryonic stem cells (ESCs), neural stem cells (NSCs), mesenchymal stem cells (MSCs), and induced pluripotent stem cells (iPSCs) [12].

PCL is a semi-crystalline and hydrophobic aliphatic polyester, biocompatible and bioresorbable. It has received a great attention for biomedical applications as an implantable biomaterial for sutures, wound dressing, and scaffolds for tissue repair [13,14]. In our previous works [15,16], flat membranes of plain PCL, PCL/GO, and PCL/partially reduced graphene oxide (rGO) were developed by a facile phase inversion fabrication method using nontoxic reagents. The membranes showed high porous morphology that provided favorable nutrient transport properties, as well as suitable cell adhesion and proliferation, particularly, PCL with graphene based nanoplatelets. Furthermore, the introduction of rGO into the PCL matrix enhanced the nutrient transport properties, which suggested the increased water wettability of the membranes. Therefore, the PCL/rGO membranes were considered to have great potential to act as scaffolds for neural cells in perfusion bioreactors, in particular, for the regeneration of neural tissue from stem cells from human origin to fabricate in vitro human neural models.

After figuring out the promising properties exhibited by the PCL/rGO membranes and accounting for their use in neural models, the study of the in vitro hydrolytic degradation route and stability behavior of these innovative membranes, as a non-permanent scaffolding material, is crucial. Ideal scaffolds should maintain their properties for sufficient time to complete their function [17]. PCL is a long-term stable polymer when subjected to hydrolytic degradation conditions, and therefore requires 2–4 years for its complete degradation, depending on the starting molecular weight of the PCL [14]. However, it has been widely observed that the incorporation of carbon based nanomaterials into the polymer matrix on the hydrolytic degradation alters the degradability of the polymeric matrix. For instance, Duan et al. [18] observed the enhancement of the wettability of a poly(L-lactide)(PLLA)-GO composite, a behavior that was ascribed to the presence of oxygen-containing groups on the surface of the GO nanoplatelets, facilitated the scission of the polymer macromolecular chains, and consequently, the scaffolds experienced a faster hydrolytic degradation. Similar behavior was observed by Zhao et al. [19], who developed nanocomposites of PLLA with multiwalled carbon nanotubes (MWNTs). Nevertheless, the literature also provides studies showing that the incorporation of graphene nanoplatelets into the polymer matrix produced the opposite effect, i.e., a reduction of the biodegradation rate of the material, due to the hydrophobic constitution of graphene [20,21]. Recently, Murray et al. [21] reported the enzymatic degradation of PCL/rGO mixtures and composites. On the one hand, the presence of rGO below 5 *w/w* % did not significantly influence the enzymatic degradation kinetics. On the other hand, rGO incorporation above 5 *w/w* % in the PCL/rGO composites caused a deceleration of the enzymatic degradation. This was attributed to the higher hydrophobicity of the composite PCL/rGO materials. While enzymatic degradation facilitates the analysis of the degradation changes on the non-permanent polymer devices under acceptable timeframes, long-term hydrolytic studies simulate physiological conditions more adequately. However, it has been demonstrated that accelerated enzymatic studies of PCL scaffolds led to very different degradation mechanisms, and consequently functional properties, than long-term hydrolytic degradation [22]. To the best of our knowledge, the long-term hydrolytic degradation of PCL/graphene composites or blends has not been reported so far.

The aim of this work consisted in the study of the long-term hydrolytic degradation of mixed matrix membranes of PCL/rGO [16]. Also, plain PCL membranes under hydrolytic degradation were evaluated and compared. In vitro conditions were simulated by immersion of the membranes in a phosphate buffer solution (PBS, pH 7.4) at 37 °C. The evolution of the functional, morphological, chemical, and thermal characteristics of the PCL/rGO membranes was evaluated during a period of 1 year. A degradation kinetics and hydrolytic pathway of the membranes were proposed and their structural stability was analyzed. Additionally, the degradation products during the study were monitored in order to elucidate potential effects on cell cytotoxicity.

2. Materials and Methods

2.1. Membrane Preparation

PCL pellets (average molecular weight, 80 kDa; Sigma-Aldrich, Madrid, Spain) were used to fabricate PCL/rGO membranes using a phase inversion technique. The synthesis of rGO particles adapted from Ribao et al. [23], as well as the fabrication of the membranes, was described in detail in our previous work [16]. Control PCL membranes without rGO nanoplatelets were also prepared for comparison.

Hydrolytic degradation experiments were performed on PCL/rGO and PCL membranes working under simulated in vitro bioreactor conditions. A sufficient number of membranes were submerged in a phosphate buffer solution (PBS, pH 7.4) and placed in an incubator at 37 °C. Separate solutions were used for testing PCL/rGO and PCL membranes. Samples were taken out of the solution for characterization at predetermined degradation time intervals: 0, 2, 4, 6, 9, and 12 months. PBS was prepared as follows: 8 g of NaCl, 0.2 g of KCl, 1.44 g of Na_2HPO_4, and 0.24 g of KH_2PO_4 were solubilized in 800 mL of distilled water. Then, the pH was adjusted to 7.4 with HCl (0.1 mol/L) and made up to 1 L with distilled water. PBS was autoclaved for sterilization. The membranes were sterilized by immersion in ethanol/water 70/30 *v/v* % and subsequent exposure to UV light for 20 min in a laminar cabinet.

2.2. Characterization

2.2.1. Functional Properties

Axial tensile tests of the membranes were done using a servo-hydraulic testing universal machine (ME-400, SERVOSIS, Madrid, Spain) following the ISO standard for thin plastic membranes (ASTM D882-12). The specimens had an area of 40×6 mm^2, and the tests were carried out using a load cell of 1.25 kN at a constant elongation speed of 8 mm/s.

A tangential flow filtration system was used to characterize the flux of nutrients across the membranes. The cross-flow filtration set up was already defined in our previous work [16]. A model feed solution was prepared, consisting of protein bovine serum albumin (BSA, >96% purity, Sigma-Aldrich) at 0.4 g/L in PBS (pH 7.4). The membrane was previously stabilized with ultrapure (UP) water at 0.1 bar for 1 h. Afterwards, the BSA model solution was circulated throughout the feed compartment of the membrane cell, and a transmembrane pressure of 0.1 bar was applied during 4 h of operation. The permeate solution was collected and weighed while the retentate was recirculated to the feed tank. The change with time (t) of total BSA solution flux (J_T (L·m^{-2}·h^{-1})) was determined as

$$J_T = \left(W_{T,permeated} \times \rho_{PBS,37°C} \right) / (\Delta t \times A_e) \tag{1}$$

where $W_{T,permeated}$ (g) is the mass of permeate collected in the time interval Δt (h) and using an effective surface area A_e (m^2), and $\rho_{PBS,37°C}$ (g·L^{-1}) the PBS density at 37 °C. At least two membrane replicates were analyzed for each degradation time.

2.2.2. Physical–Chemical Properties

The average molecular weight of the membranes was determined by gel permeation chromatography (GPC model 510, Waters, Madrid, Spain). Three size exclusion chromatography columns of styrene divinyl benzene copolymer were placed in series (model Styragel HR 5E, Waters) and a refractometer (model 410, Waters) was used for detection. The columns were thermostatized at 40 °C, and the measurements were carried out using 1 mL/min of tetrahydrofuran (THF 99.9%, Panreac, Barcelona, Spain) as carrier. Membrane samples were solubilized in THF at a concentration of 0.5 mg/mL. PCL/rGO in THF samples were centrifuged for 1 h and filtered through a 0.45 μm filter before GPC injection, in order to avoid rGO contamination of the GPC columns. The values of molecular weight distribution were obtained by the Empower 2 software (Waters). The molecular weights were determined using a universal calibration curve related to polystyrene standards (Shodex, Waters, Cerdanyola del Vallès, Spain) corrected by the Mark–Houwink–Sakura equation and the corresponding PCL coefficients. Measurements were done in duplicate.

Thermal properties of the samples at 0 and 12 months of degradation were evaluated by differential scanning calorimetry (DSC, DSC-131, SETARAM Instrumentation, Caluire, France) at a scan rate of 10 °C/min. Samples (5–10 mg) were heated from room temperature to 100 °C (first heating run). After 10 min stabilization at 100 °C, the samples were cooled down to 0 °C (cooling run) and finally heated up again to 100 °C (second heating run) after stabilization. The degree of crystallinity, χ_C (%) was calculated using Equation (2) [24], where ΔH_m^0 (J·g^{-1}) is the melting enthalpy calculated from the second heating ramp, ΔH_m^0 is the melting enthalpy for a 100% crystalline PCL (139.5 J·g^{-1} [24]) and is the mass fraction of rGO in the PCL membrane.

$$\chi_C = \Delta H_m / \left[(1 - \beta) \times \Delta H_m^0 \right] \qquad (2)$$

The concentration of 6-hydroxycaproic acid (6-HCA), typically found as monomer degradation product of PCL, was analyzed in the PBS medium where the membranes were submerged. The UV–vis spectrophotometer (UV-1800 model, Shimadzu, Duisburg, Germany) was set at a 210 nm wavelength [25]. Measurements of 6-HCA were carried out after 6 and 12 months of degradation time. The presence of rGO nanoplatelets on the PBS medium and in the membrane matrix after 12 months of degradation was analyzed (see Supplementary Materials).

Microscopic images of the membranes were obtained using a scanning electron microscope (SEM, EVO MA 15, Carl Zeiss, Madrid, Spain) at a voltage of 20 kV, in order to determine the structure and morphology of the surface and cross section of the membranes. Samples for the cross-section images were frozen in liquid nitrogen to be fractured. All the samples were kept overnight at 30 °C under vacuum, and were gold sputtered before examination. Moreover, visual inspection of the membrane was recorded by taking photographs of the same membrane specimen periodically.

Before any testing, membranes were cleaned with UP water to remove any possible salt deposit. Results are expressed as average ± standard deviation.

3. Results

3.1. Functional Properties

Figure 1 shows the mechanical properties of PCL and PCL/rGO membranes during the degradation study. Figure 1A–D shows the Young modulus, yield point, ultimate tensile strength, and ultimate strain, respectively. Mechanical tests were not feasible beyond 4 months, due to the loss of mechanical stability. Specifically, the PCL membranes could be handled until 6 months, while the PCL/rGO membranes could not be manipulated after 4 months. Videos confirming the poor stability of the PCL and PCL/rGO membranes after 12 months of immersion in PBS are included in the Supplementary Materials (Videos S1 and S2). During the hydrolytic degradation study, a gradual reduction in the mechanical properties of both membranes was observed in the values of mechanical parameters (Figure 1). Overall, at time 0, the presence of rGO in the polymer matrix significantly reduced the mechanical properties in comparison to the plain PCL membranes. After 2 months, PCL and PCL/rGO membranes showed homogeneous reduction of mechanical properties. For instance, PCL membranes showed a reduction of mechanical properties in the range 57–62%, in a narrow range for all the properties evaluated. The reduction of properties for PCL/rGO membranes was also encountered, mainly around 63–68%, with the exception of the Young Modulus that suffered a 41% drop. After 4 months, PCL/rGO still suffered similar reduction of mechanical properties as in previous degradation times, while PCL membranes presented more disordered behavior: for instance, the yield point barely changed while the Young modulus showed a 60% reduction.

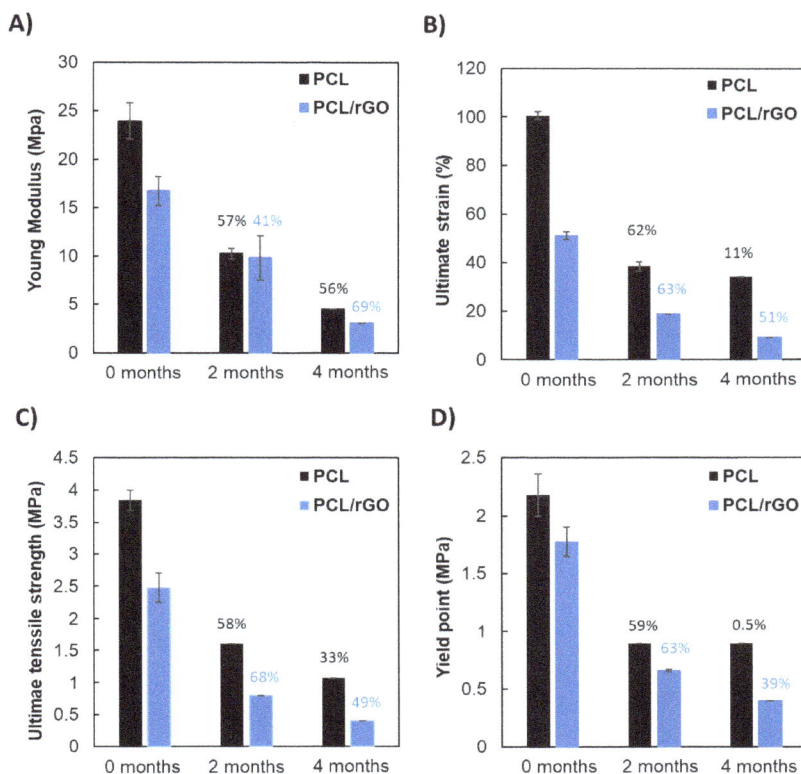

Figure 1. Mechanical properties of poly(ε-caprolactone) (PCL) and PCL/reduced graphene oxide (rGO) membranes. (**A**) ultimate tensile stress; (**B**) Young modulus; (**C**) ultimate strain; and (**D**) yield point for PCL and PCL/rGO membranes at 0, 2, and 4 months of degradation. (% values represent the reduction of the mechanical parameters between degradation times.)

Figure 2 plots the change with the filtration time of the volumetric flux of BSA model solution through the PCL membranes for specimens that had been submerged in the hydrolytic bath for 0, 2, 4, and 6 months, and PCL/rGO only at $t = 0$ months. BSA permeation tests for PCL/rGO membranes after in vitro degradation ($t > 0$ months) could not be performed because they could not withstand the transmembrane pressure of the filtration device. The membranes experienced a sharp flux drop during the first 2 h of filtration, with a reduction of $88.1 \pm 2.9\%$ in each point of degradation. Afterwards, it could be assumed that the flux reached a pseudo steady state (Inset of Figure 2). The BSA solution fluxes at this steady state were as follows: $143 \pm 66 > 108 \pm 5 > 103 \pm 3 > 80 \pm 7$ L·m^{-2}·h^{-1} at 0, 2, 4, and 6 months of degradation, respectively. Similarly, the BSA solution flux through PCL/rGO membranes at 0 months decayed from the initial value of 3620 ± 356 L·m^{-2}·h^{-1} to a stable flux of 190 ± 68 L·m^{-2}·h^{-1} (drop of $94.5 \pm 2.4\%$).

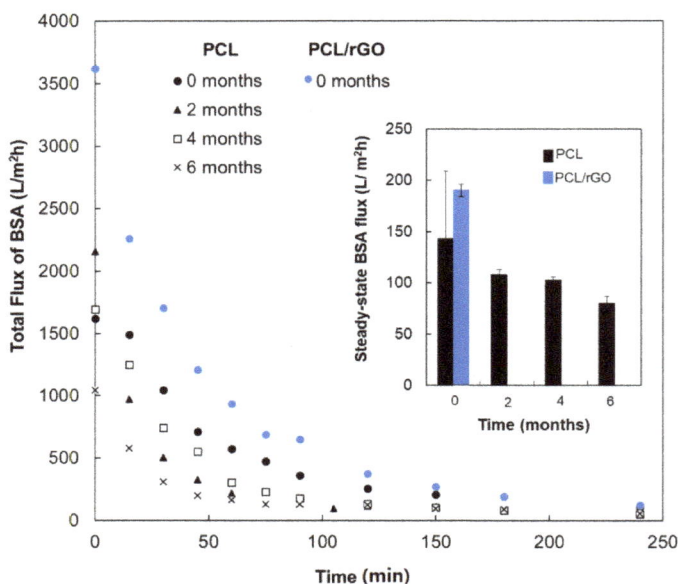

Figure 2. Total flux decay of BSA model solution of average values of PCL membranes during 240 min at different times of degradation. Flux data for PCL/rGO membrane at t = 0 months were also included (deviation bars not shown for the shake of clarity). Inset shows the values of BSA solution flux at steady state for PCL and PCL/rGO membranes at 0, 2, 4, and 6 months of degradation.

3.2. Physical–Chemical Properties Characterization

Figure 3A shows the hydrolysis degradation pathway of the PCL polymer. The cleavage of the ester bonds of PCL is produced upon the reaction with water, forming carboxyl end-groups, and the progressive reduction of the average molecular size to give water-soluble degradation products, including oligomers and monomer (6-HCA), that diffused out of the membrane matrix and solubilized in the PBS medium [26].

The progress of the number average molecular weight (M_n) with the degradation time is presented in Figure 3B. Both PCL and PCL/rGO membranes showed a progressive decrease in M_n. After two months of degradation, the M_n of the membranes suffered a significant reduction from the initial value of 75 ± 6 kDa to 61 ± 7 kDa for PCL membranes (drop of 19%), and to 49 kDa for PCL/rGO membranes (drop of 35%). At 12 months, M_n decreased further to 33 ± 0.04 kDa (56%) for PCL membranes, and to 27 ± 0.75 kDa (65%) for PCL/rGO membranes. The polydispersity index (PDI) of the molecular weight distribution remained almost constant during the degradation period, i.e., PDI values of 1.42 at t = 0 months and 1.47 at t = 12 months for PCL films, and 1.34 at t = 12 months for PCL/rGO films were obtained (Figure 3B). Regarding the hydrolysis kinetics of polymers, it usually follows second order reaction kinetics, that is, the rate of the reaction is proportional to the concentration of water and the concentration of chemical bonds susceptible of hydrolysis, i.e., carboxylic bonds for polyesters [27] (see Equations (S1) and (S2) in Supplementary Materials). Figure S1 (Supplementary Materials) shows good agreement of the fitting of the experimental data to $1/M_n$ vs t, which indicates that the hydrolysis of our membranes proceeded according to second order kinetics. Moreover, in Figure S1, it was shown that the hydrolysis kinetics of PCL/rGO membranes was only slightly faster (approximately 1.4 times) than the hydrolytic kinetics of PCL membranes.

A)

Poly(ε-caprolactone) (PCL) Hydrolysis intermediates Monomer 6-hydroxicaproic acid (6-HCA)

B)

C)

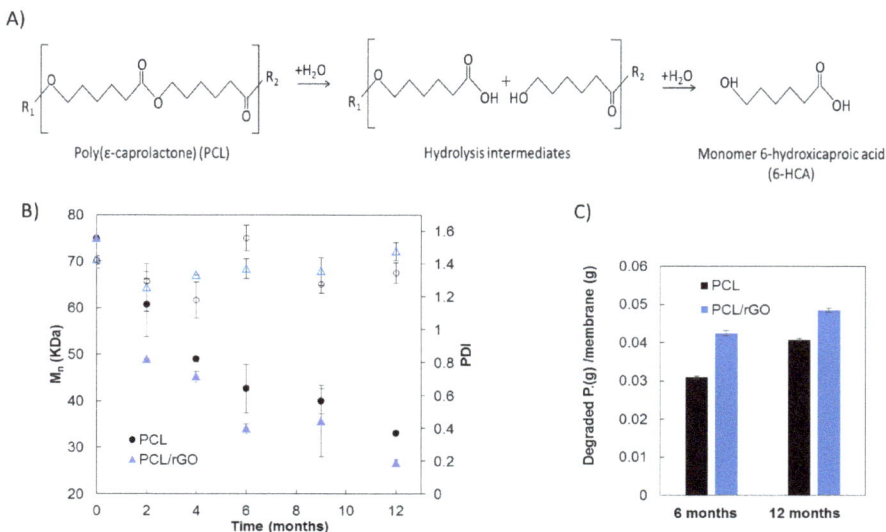

Figure 3. (**A**) Scheme of the PCL hydrolytic degradation process, adapted from Woodruff et al. [14]; (**B**) Change of the number average molecular weight (M_n, filled symbols) and polydispersity index (PDI, empty symbols); and (**C**) mass of degradation product 6-HCA in the PBS formed during degradation process of PCL and PCL/rGO membranes.

Figure 3C illustrates the concentration of 6-HCA per unit mass of membrane released to the PBS medium after 6 and 12 months of hydrolytic degradation. The concentration of the 6-HCA in the PBS increased significantly with the degradation time, as expected. The 6-HCA concentration in the PBS medium was higher for PCL/rGO than for PCL membranes, in good agreement with the evolution of the molecular size observed in Figure 3B. The presence of 6-HCA in the buffer solutions barely affected the pH (data not shown). During the degradation test, an attempt to evaluate what happened with the rGO nanoplatelets of the PCL/rGO membranes was done. After re-dissolving the PCL/rGO membranes in THF, the presence of rGO nanoplatelets in the membranes was visually confirmed at 0 months and after 12 months of degradation (see Figure S2 of the Supplementary Materials). It can also be appreciated that the precipitated rGO was qualitatively more abundant (higher mass concentration) in the samples corresponding to 12 months of degradation than in the PCL/rGO membranes at *t* = 0. Moreover, the UV–vis spectrum of the PBS that contained the PCL/rGO membranes during 12 months of degradation did not show the rGO representative peak around 270 nm [28]. These qualitative results led us to think that rGO mainly remained in the solid material of the PCL/rGO membranes.

Furthermore, Figure 4 represents the DSC thermograms of the PCL (A) and PCL/rGO membranes (B) at 0 and 12 months. The initial value of the melting temperature (T_m) (0 months) was 62.10 °C for PCL membranes and 60.36 °C for PCL/rGO membranes. After 12 months of degradation time, T_m increased to 64.54 °C for PCL and to 64.72 °C for PCL/rGO membranes. The crystallization temperature, T_c, increased as well during the degradation period in both types of membranes, from 31.75 °C to 32.63 °C for PCL, and from 32.45 °C to 35.00 °C for PCL/rGO. The initial χ_C of PCL/rGO was 41%, and increased to 46% after 12 months. Meanwhile, PCL membranes crystallinity varied from 35% to 44%. The higher crystallinity of the degraded samples pointed to the preferential hydrolytic attack of the amorphous polymer phase.

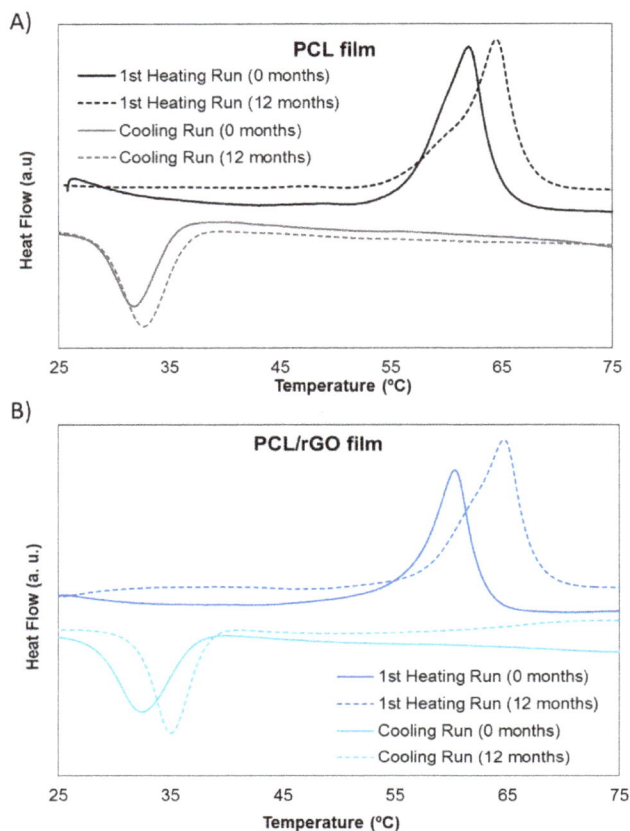

Figure 4. DSC thermogram of PCL (**A**) and PCL/rGO (**B**); membranes at 0 and 12 months of degradation.

The membranes did not suffer any significant reduction in dimensions (width, length, and thickness) during the degradation time (Figure S3, Supplementary Materials). Regarding the microscopic morphology, SEM images of the surface and cross section of the PCL and PCL/rGO membranes at 0, 2, and 12 months of hydrolytic degradation are illustrated in Figure 5. Overall, a noticeable change in the morphology of the membranes can be observed. The surface of both PCL and PCL/rGO membranes eroded slightly from 0 to 2 months, and then very notably after 12 months of degradation. The morphological degradation of the internal structure of the membranes was also evident, as shown in the SEM cross section images.

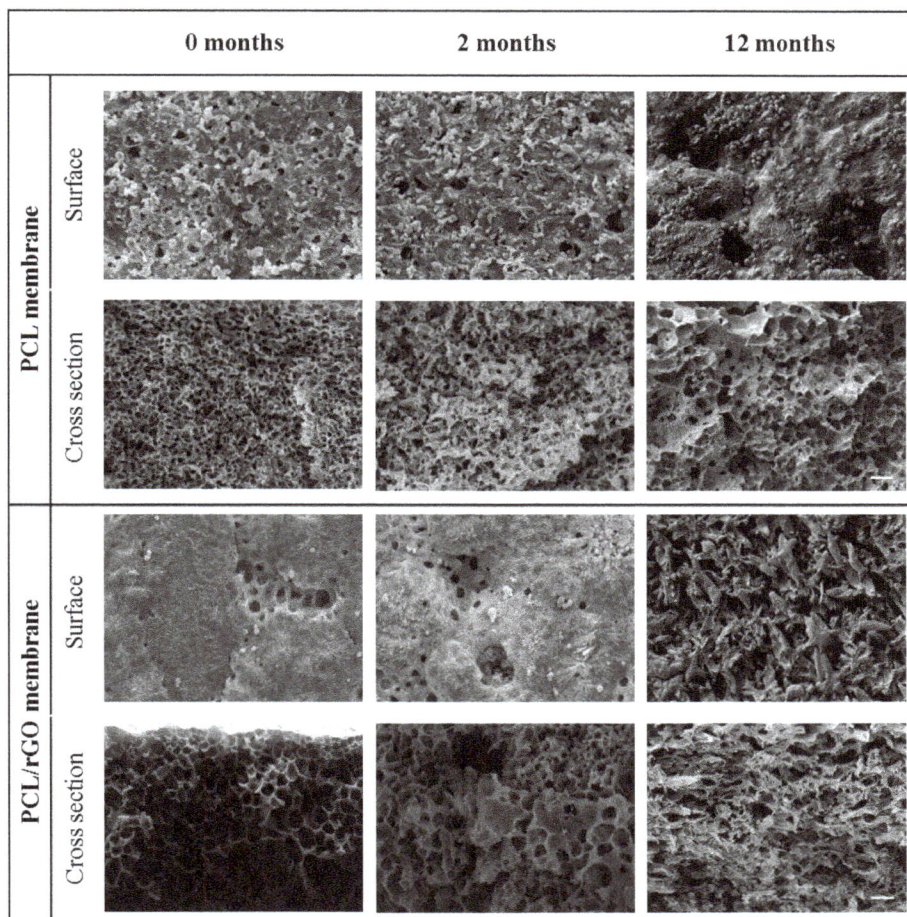

Figure 5. SEM images of PCL and PCL/rGO membranes at 0, 2, and 12 months of hydrolytic degradation. The scale bars represent 10 μm.

4. Discussion

PCL/rGO and control PCL membranes fabricated in this work degraded continuously under the presence of PBS simulating in vitro culture conditions. The molecular weight presented a progressive reduction (Figure 3B), produced by the hydrolytic chain scission of the ester group due to water penetration. The degradation kinetics of our membranes corresponded to second order kinetics, in agreement with typical hydrolysis of large molecular weight polyesters [27]. Also, the maintenance of PDI values (Figure 3B) during the degradation period indicated that all the carboxylic bonds of the polymer chain had equal reactivity, in agreement with the obtained second order kinetics. In spite of the hydrophobic character of the rGO nanoplatelets [10], a slight, though not significant, acceleration of the hydrolytic degradation for PCL/rGO membrane was observed in comparison to the plain PCL membranes in terms of molecular weight change. The monomer 6-HCA, as the main indicator of degradation products, was released and diffused into the buffer media (Figure 3C). Therefore, its concentration progressively increased with the degradation period, and showed slightly higher values for buffer media containing PCL/rGO membranes than for PCL membranes, in agreement with the results of the molecular weight degradation kinetics. Moreover, the higher degree of crystallinity, as well as the increase on the thermal properties after 12 months of degradation, pointed to the preferential hydrolytic attack of the polymer amorphous region in both membranes [17]. Finally, the internal morphology of the membranes suffered a clear change (Figure 5), while the dimensions of the membranes remained constant during the degradation process (Figure S3). According to the aforementioned results, in the present work, degradation of the membranes proceeded via bulk degradation mechanism [29]. Bulk degradation mechanism of PCL networks was also reported by [22] under similar hydrolytic degradation conditions. The similar tendencies observed in the present work on the molecular weight reduction for PCL and PCL/rGO membranes, points to the high porosity of the fabricated membranes as the main cause for the bulk hydrolysis mechanism. The porous internal morphology favored the water penetration and the outward diffusion of the degradation products [30]. In comparison to other reported works with similar molecular weight (PCL 80 kDa) [31,32], our fabricated membranes demonstrated an accelerated degradation rate.

All mid- and end-point degradation products must be thoroughly investigated for possible immunogenic reactions [33]. During the progress of the polymer degradation, it was observed that the rGO nanoplatelets remained mainly in the membrane. This was consistent with the results of Murray et al. [21]. They also reported that PCL/rGO blended materials increased the relative concentration of rGO during enzymatic degradation from 5 *w/w* % to 19 *w/w* %, and did not observe cytotoxicity on L-929 fibroblast cells growing for short periods. In our previous work [16], we also observed a positive biocompatibility on glioblastoma cells of the PCL membranes containing rGO nanomaterials after 14 days of culture. These null cytotoxic results are also in agreement with the low pH acidification of the buffer solution observed in our system (results not shown) that would not likely turn into a negative cellular response [34].

The presence of rGO nanoplatelets in the polymer matrix significantly reduced the mechanical integrity of the membranes at any degradation time. This effect was attributed to a restriction of the mobility of the polymer chains [24], and to defects and gaps created by the presence of rGO in the polymer matrix [35]. Also, the presence of rGO caused a faster and more intense loss of mechanical properties and structural stability for PCL/rGO membranes in contrast to PCL membranes (Figure 1 and Videos S1 and S2) that could also be explained by the same causes that decreased the initial mechanical properties of PCL/rGO in contrast to PCL membranes, as previously explained. Despite the fast loss of mechanical properties of the PCL/rGO membranes, after 4 months of degradation, these materials still comply sufficiently with the mechanical properties required for materials to sustain neural tissues. Actually, the mechanical stiffness of the 4-month degraded membranes are closer to the values of the hydrogels typically employed as scaffold materials for neural tissue regeneration, i.e., Matrigel [36], modified gelatin [37], polyethylene glycol, or alginate hydrogels [38,39]. For instance, a broad range of Young modulus values, i.e., in the order of 0.2–20 kPa for alginate hydrogel 40 and 0.1–1.2 MPa for modified gelatin [37], can be found for these materials. Nevertheless, although Matrigel is one of the most employed materials for scaffolds in neural adhesion and proliferation, its relatively weak mechanical strength and significant degradation over long-term culture has been considered a drawback for its use in in vitro neural models [40]. Apart from the effect of the membrane stiffness on the induction of mechanical cues over the cells, it is envisaged that the mechanical properties loss of the material could be substituted by the tissue mechanical stability if there is an equivalent rate of membrane structural disintegration and tissue regeneration. In the field of neural tissue regeneration, in vitro models of cerebral organoids require only 8–10 days for the appearance of neural identity and 20–30 days for the formation of defined brain regions [36]. Mahoney and Anseth [39] confirmed the suitable use of polyethylene glycol hydrogels to act as cell carriers for transplantation into the central nervous system (CNS), with an accelerated loss of mechanical properties in 12 days. In general, an adequate scaffold material should lose mechanical properties at an approximate rate of 8%/week during in vivo degradation [41], and neural scaffold materials would ideally degrade over a period of 2–8 weeks via hydrolysis, ion exchange, or through enzymatic reactions [33]. All the previous works support the idea that the rate of neural tissue regeneration could be comparable to the degradation rate behavior of our PCL and mainly to PCL/rGO membranes. Actually, preliminary experiments on neural progenitor cells (NPC) differentiation and maturation have been performed for 20 days on PCL and PCL/rGO membranes, showing promising cell coverage (unpublished data) and adequate structural integrity for the manipulation.

Regarding the behavior of the BSA solution flux through the membranes, the initial pronounced reduction of the flux (Figure 2) was associated with the internal fouling due to BSA protein adhesion to the pore walls [16]. The progressive decrease of the steady-state BSA fluxes for PCL membranes during the degradation time could be attributed to the gradual loss of structural integrity under hydrodynamic pressure, causing the membrane compaction, and therefore, the pore size reduction during filtration assays [30]. Regardless, the significant reduction of the nutrient flux at steady state, PCL and PCL/rGO membranes still displayed a comparable total BSA solution flux to that reported by Bettahalli et al. [42] for commercial poly(ether-sulfone) hollow fibers, theoretically sufficient to supply the needs of glucose consumption to more than three layers of cells under confluence in a perfusion bioreactor.

5. Conclusions

The present work reports on the evaluation of the hydrolytic degradation of novel PCL/rGO porous membranes fabricated by phase inversion technique. The hydrolytic degradation during a long term period of 12 months of these PCL/rGO membranes was evaluated in this work, in order to study the membrane capacity to act as scaffold for in in vitro bioreactors for neural tissue regeneration and its further use as in vitro human neural models.

Both, PLC/rGO membranes and PCL membranes (control membranes) exhibited a fast degradation rate. This work demonstrates that the high porous membrane structure obtained as a result of the phase inversion manufacturing technique was the main factor on the acceleration of the degradation, as it could promote water penetration, and therefore facilitate the bulk hydrolytic mechanism of the membranes. The molecular weight decreased, following second order kinetic rate, characteristic of these types of polyesters of large molecular weight. As a result, there was a loss of the membrane's mechanical resistance, an enhancement of the crystallinity, and the formation of PCL degradation products, such as the monomer 6-hydroxycaproic acid, released to the hydrolytic media. Besides the aforementioned alterations, the changes in the porous morphology without any observable modification of the sample dimensions led to the conclusion that degradation proceeded via bulk hydrolysis mechanism. The introduction of rGO nanoplatelets into the PCL matrix only slightly accelerated the degradation rate. Particularly, the presence of rGO reduced significantly the mechanical stability of the membranes at all degradation times. However, PCL/rGO membranes still procured sufficient mechanical properties to theoretically comply with the specifications of the neural tissue regeneration. Besides, the degradation rate of the membranes herein reported would perfectly fit the rate of neural tissue regeneration that would need around 1 month to be completed. The rGO nanoplatelets remained preferentially in the polymer matrix of the membrane during the degradation process and, according to previous works, the degradation products of similar PCL/rGO blended materials should not alter the cytotoxicity of the buffer solution. The high porosity that induces exceptional BSA solution flux let us deem that PCL/rGO membranes would be promising candidates to be used as scaffolds for neural tissue regeneration in perfusion bioreactors. Finally, it has to be remarked that experiments to evaluate the performance of the PCL/rGO membranes on dynamic neural cell culture, as well as the assessment of the potential of PCL/rGO membranes to induce stem cell differentiation into neural tissue, are currently under progress.

Supplementary Materials: The following are available online at www.mdpi.com/2077-0375/8/12/s1, Video S1: Mechanical behavior of PCL membranes after 12 months of hydrolytic degradation; Video S2: Mechanical behavior of PCL/rGO membranes after 12 months of hydrolytic degradation; Equations (S1) and (S2) demonstrating simplifications and fitting of the experimental data of change of the polymer molecular weight with degradation time to a 2nd order hydrolysis kinetics and Figure S1: Kinetics of the hydrolysis of PCL and PCL/rGO membranes. Fitting of the molecular weight to the second order hydrolysis kinetics (Equation (S2)) is depicted (dotted lines). Figure S2: Photographs showing the rGO content at 0 and 12 months of degradation of the PCL/rGO membranes. Figure S3: Photographs showing the visual aspect of the wet PCL and PCL/rGO membranes at 0 and 12 months of degradation.

Acknowledgments: Financial support of the Cantabria Explora call through project JP03.640.69 is gratefully acknowledged. The support of project CTM2016-75509-R (MINECO and FEDER-Spain) is granted. We also thank Marta Romay at University of Cantabria who performed part of the experiments.

Author Contributions: N.D. conceived and designed the experiments; S.S.-G. performed the experiments; S.S.-G., N.D and A.U. analyzed the data; S.S.-G., N.D. and A.U. wrote the paper.

Conflicts of Interest: The authors declare no conflict of interest.

References

1. Guo, B.; Lei, B.; Li, P.; Ma, P.X. Functionalized scaffolds to enhance tissue regeneration. *Regen. Biomater.* **2015**, *2*, 47–57. [CrossRef] [PubMed]

2. Liu, X.; Holzwarth, J.M.; Ma, P.X. Functionalized Synthetic Biodegradable Polymer Scaffolds for Tissue Engineering. *Macromol. Biosci.* **2012**, *12*, 911–999. [CrossRef] [PubMed]

3. Kim, Y.H.; Jyoti, M.A.; Song, H.Y. Immobilization of cross linked Col-I-OPN bone matrix protein on aminolysed PCL surfaces enhances initial biocompatibility of human adipogenic mesenchymal stem cells (hADMSC). *Appl. Surf. Sci.* **2014**, *303*, 97–106. [CrossRef]

4. Wang, Z.; Sun, B.; Zhang, M.; Ou, L.; Che, Y.; Zhang, J.; Kong, D. Functionalization of electrospun poly(ε-caprolactone) scaffold with heparin and vascular endothelial growth factors for potential application as vascular grafts. *J. Bioact. Compat. Polym.* **2012**, *28*, 154–166. [CrossRef]

5. Patrício, T.; Domingos, M.; Gloria, A.; Bártolo, P. Characterisation of PCL and PCL/PLA scaffolds for tissue engineering. *Procedia CIRP* **2013**, *5*, 110–114. [CrossRef]

6. Pêgo, A.P.; Poot, A.A.; Grijpma, D.W.; Feijen, J. Copolymers of trimethylene carbonate and epsilon-caprolactone for porous nerve guides: Synthesis and properties. *J. Biomater. Sci. Polym. Ed.* **2001**, *12*, 35–53. [CrossRef] [PubMed]

7. Koupaei, N.; Karkhaneh, A. Porous crosslinked polycaprolactone hydroxyapatite networks for bone tissue engineering. *Tissue Eng. Regen. Med.* **2016**, *13*, 251–260. [CrossRef]

8. Augustine, R.; Dominic, E.A.; Reju, I.; Kaimal, B.; Kalarikkal, N.; Thomas, S. Electrospun polycaprolactone membranes incorporated with ZnO nanoparticles as skin substitutes with enhanced fibroblast proliferation and wound healing. *RSC Adv.* **2014**, *4*, 24777–24785. [CrossRef]

9. Crowder, S.W.; Liang, Y.; Rath, R.; Park, A.M.; Maltais, S.; Pintauro, P.N.; Hofmeister, W.; Lim, C.C.; Wang, X.; Sung, H.-J. Poly (ε-caprolactone)-carbon nanotube composite scaffolds for enhanced cardiac differentiation of human mesenchymal stem cells. *Nanomedicine* **2013**, *8*, 1763–1776. [CrossRef] [PubMed]

10. Goenka, S.; Sant, V.; Sant, S. Graphene-based nanomaterials for drug delivery and tissue engineering. *J. Control. Release* **2014**, *173*, 75–88. [CrossRef] [PubMed]

11. Shin, S.R.; Li, Y.-C.; Jang, H.L.; Khoshakhlagh, P.; Akbari, M.; Nasajpour, A.; Zhang, Y.S.; Tamayol, A.; Khademhosseini, A. Graphene-based materials for tissue engineering. *Adv. Drug Deliv. Rev.* **2016**, *105*, 255–274. [CrossRef] [PubMed]

12. Bressan, E.; Ferroni, L.; Gardin, C.; Sbricoli, L.; Gobbato, L.; Ludovichetti, F.; Tocco, I.; Carraro, A.; Piattelli, A.; Zavan, B. Graphene based scaffolds effects on stem cells commitment. *J. Transl. Med.* **2014**, *12*, 296. [CrossRef] [PubMed]

13. Mondal, D.; Griffith, M.; Venkatraman, S.S. Polycaprolactone-based biomaterials for tissue engineering and drug delivery: Current scenario and challenges. *Int. J. Polym. Mater. Polym. Biomater.* **2016**, *65*, 255–265. [CrossRef]

14. Woodruff, M.A.; Hutmacher, D.W. The return of a forgotten polymer—Polycaprolactone in the 21st century. *Prog. Polym. Sci.* **2010**, *35*, 1217–1256. [CrossRef]

15. Diban, N.; Ramos-Vivas, J.; Remuzgo-Martinez, S.; Ortiz, I.; Urtiaga, A. Poly (ε-caprolactone) films with favourable properties for neural cell growth. *Curr. Top. Med. Chem.* **2014**, *14*, 2743–2749. [CrossRef] [PubMed]
16. Diban, N.; Sanchez-Gonzalez, S.; Lázaro-Díez, M.; Ramos-Vivas, J.; Urtiaga, A. Facile fabrication of poly(ε-caprolactone)/graphene oxide membranes for bioreactors in tissue engineering. *J. Membr. Sci.* **2017**, *540*, 219–228. [CrossRef]
17. Bosworth, L.A.; Downes, S. Physicochemical characterisation of degrading polycaprolactone scaffolds. *Polym. Degrad. Stab.* **2010**, *95*, 2269–2276. [CrossRef]
18. Duan, J.; Xie, Y.; Yang, J.; Huang, T.; Zhang, N.; Wang, Y.; Zhang, J. Graphene oxide induced hydrolytic degradation behavior changes of poly(L-lactide) in different mediums. *Polym. Test.* **2016**, *56*, 220–228. [CrossRef]
19. Zhao, Y.; Qiu, Z.; Yang, W. Effect of multi-walled carbon nanotubes on the crystallization and hydrolytic degradation of biodegradable poly(l-lactide). *Compos. Sci. Technol.* **2009**, *69*, 627–632. [CrossRef]
20. Finniss, A.; Agarwal, S.; Gupta, R. Retarding hydrolytic degradation of polylactic acid: Effect of induced crystallinity and graphene addition. *J. Appl. Polym. Sci.* **2016**, *133*, 1–8. [CrossRef]
21. Murray, E.; Thompson, B.C.; Sayyar, S.; Wallace, G.G. Enzymatic degradation of graphene/polycaprolactone materials for tissue engineering. *Polym. Degrad. Stab.* **2015**, *111*, 71–77. [CrossRef]
22. Castilla-Cortázar, I.; Más-Estellés, J.; Meseguer-Dueñas, J.M.; Escobar Ivirico, J.L.; Marí, B.; Vidaurre, A. Hydrolytic and enzymatic degradation of a poly(ε-caprolactone) network. *Polym. Degrad. Stab.* **2012**, *97*, 1241–1248. [CrossRef]
23. Ribao, P.; Rivero, M.J.; Ortiz, I. TiO$_2$ structures doped with noble metals and/or graphene oxide to improve the photocatalytic degradation of dichloroacetic acid. *Environ. Sci. Pollut. Res.* **2016**, *24*, 12628–12637. [CrossRef] [PubMed]
24. Wang, G.S.; Wei, Z.Y.; Sang, L.; Chen, G.Y.; Zhang, W.X.; Dong, X.F.; Qi, M. Morphology, crystallization and mechanical properties of poly(e-caprolactone)/graphene oxide nanocomposites. *Chin. J. Polym. Sci.* **2013**, *31*, 1148–1160. [CrossRef]
25. Hafeman, A.E.; Zienkiewicz, K.J.; Zachman, A.L.; Sung, H.-J.; Nanney, L.B.; Davidson, J.M.; Guelcher, S.A. Characterization of the Degradation Mechanisms of Lysine-derived Aliphatic Poly(ester urethane) Scaffolds. *Biomaterials* **2011**, *32*, 419–429. [CrossRef] [PubMed]
26. Bölgen, N.; Menceloglu, Y.Z.; Acatay, K.; Vargel, I.; Piskin, E. In vitro and in vivo degradation of non-woven materials made of poly(ε-caprolactone) nanofibers prepared by electrospinning under different conditions. *J. Biomater. Sci. Polym. Ed.* **2005**, *16*, 1537–1555. [CrossRef] [PubMed]
27. Lyu, S.; Untereker, D. Degradability of polymers for implantable biomedical devices. *Int. J. Mol. Sci.* **2009**, *10*, 4033–4065. [CrossRef] [PubMed]
28. Krishnamoorthy, K.; Veerapandian, M.; Zhang, L.H.; Yun, K.; Kim, S.J. Antibacterial efficiency of graphene nanosheets against pathogenic bacteria via lipid peroxidation. *J. Phys. Chem. C* **2012**, *116*, 17280–17287. [CrossRef]
29. Von Burkersroda, F.; Schedl, L.; Göpferich, A. Why degradable polymers undergo surface erosion or bulk erosion. *Biomaterials* **2002**, *23*, 4221–4231. [CrossRef]
30. Zhang, Q.; Jiang, Y.; Zhang, Y.; Ye, Z.; Tan, W.; Lang, M. Effect of porosity on long-term degradation of poly (ε-caprolactone) scaffolds and their cellular response. *Polym. Degrad. Stab.* **2013**, *98*, 209–218. [CrossRef]
31. Höglund, A.; Hakkarainen, M.; Albertsson, A.-C. Degradation profile of poly(ε-caprolactone)—The influence of macroscopic and macromolecular biomaterial design. *J. Macromol. Sci. Part A Pure Appl. Chem.* **2007**, *44*, 1041–1046. [CrossRef]
32. Lam, C.X.F.; Savalani, M.M.; Teoh, S.-H.; Hutmacher, D.W. Dynamics of in vitro polymer degradation of polycaprolactone-based scaffolds: Accelerated versus simulated physiological conditions. *Biomed. Mater.* **2008**, *3*, 034108. [CrossRef] [PubMed]
33. Thomas, M.; Willerth, S.M. 3-D Bioprinting of Neural Tissue for Applications in Cell Therapy and Drug Screening. *Front. Bioeng. Biotechnol.* **2017**, *5*, 69. [CrossRef] [PubMed]
34. Göpferich, A. Mechanisms of polymer degradation and erosion. *Biomaterials* **1996**, *17*, 103–114. [CrossRef]
35. Jin, T.X.; Liu, C.; Zhou, M.; Chai, S.G.; Chen, F.; Fu, Q. Crystallization, mechanical performance and hydrolytic degradation of poly(butylene succinate)/graphene oxide nanocomposites obtained via in situ polymerization. *Compos. Part A* **2015**, *68*, 193–201. [CrossRef]

36. Lancaster, M.A.; Renner, M.; Martin, C.-A.; Wenzel, D.; Bicknell, L.; Hurles, M.; Homfray, T.; Penninger, J.M.; Jackson, A.P.; Knoblich, J.A. Cerebral organoids model human brain development and microcephaly. *Nature* **2013**, *501*, 373–379. [CrossRef] [PubMed]

37. Binan, L.; Tendey, C.; De Crescenzo, G.; El Ayoubi, R.; Ajji, A.; Jolicoeur, M. Differentiation of neuronal stem cells into motor neurons using electrospun poly-l-lactic acid/gelatin scaffold. *Biomaterials.* **2014**, *35*, 664–674. [CrossRef] [PubMed]

38. Banerjee, A.; Arha, M.; Choudhary, S.; Ashton, R.S.; Bhatia, S.R.; Schaffer, D.V.; Kane, R.S. The influence of hydrogel modulus on the proliferation and differentiation of encapsulated neural stem cells. *Biomaterials* **2009**, *30*, 4695–4699. [CrossRef] [PubMed]

39. Mahoney, M.J.; Anseth, K.S. Three-dimensional growth and function of neural tissue in degradable polyethylene glycol hydrogels. *Biomaterials* **2006**, *27*, 2265–2274. [CrossRef] [PubMed]

40. Wan, X.; Ball, S.; Willenbrock, F.; Yeh, S.; Vlahov, N.; Koennig, D.; Green, M.; Brown, G.; Jeyaretna, S.; Li, Z.; et al. Perfused Three-dimensional Organotypic Culture of Human Cancer Cells for Therapeutic Evaluation. *Sci. Rep.* **2017**, *7*, 9408. [CrossRef] [PubMed]

41. Wang, Y.; Kim, Y.M.; Langer, R. In vivo degradation characteristics of poly(glycerol sebacate). *J. Biomed. Mater. Res. Part A* **2003**, *66*, 192–197. [CrossRef] [PubMed]

42. Bettahalli, N.M.S.; Vicente, J.; Moroni, L.; Higuera, G.A.; Van Blitterswijk, C.A.; Wessling, M.; Stamatialis, D.F. Integration of hollow fiber membranes improves nutrient supply in three-dimensional tissue constructs. *Acta Biomater.* **2011**, *7*, 3312–3324. [CrossRef] [PubMed]

membranes **MDPI**

Article

CO$_2$ Separation in Nanocomposite Membranes by the Addition of Amidine and Lactamide Functionalized POSS® Nanoparticles into a PVA Layer

Gabriel Guerrero [1], May-Britt Hägg [1,*], Christian Simon [2], Thijs Peters [2], Nicolas Rival [2] and Christelle Denonville [2]

[1] Department of Chemical Engineering, Norwegian University of Science and Technology, 7491 Trondheim, Norway; gabriel.g.heredia@ntnu.no
[2] SINTEF Industry, 0314 Oslo, Norway; christian.r.simon@sintef.no (C.S.); Thijs.peters@sintef.no (T.P.); nicolas.rival@sintef.no (N.R.); christelle.denonville@sintef.no (C.D.)
* Correspondence: may-britt.hagg@ntnu.no; Tel.: +47-7359-4033 or +47-9308-0834

Received: 8 May 2018; Accepted: 6 June 2018; Published: 8 June 2018

Abstract: In this article, we studied two different types of polyhedral oligomeric silsesquioxanes (POSS®) functionalized nanoparticles as additives for nanocomposite membranes for CO$_2$ separation. One with amidine functionalization (Amidino POSS®) and the second with amine and lactamide groups functionalization (Lactamide POSS®). Composite membranes were produced by casting a polyvinyl alcohol (PVA) layer, containing either amidine or lactamide functionalized POSS® nanoparticles, on a polysulfone (PSf) porous support. FTIR characterization shows a good compatibility between the nanoparticles and the polymer. Differential scanning calorimetry (DSC) and the dynamic mechanical analysis (DMA) show an increment of the crystalline regions. Both the degree of crystallinity (X_c) and the alpha star transition, associated with the slippage between crystallites, increase with the content of nanoparticles in the PVA selective layer. These crystalline regions were affected by the conformation of the polymer chains, decreasing the gas separation performance. Moreover, lactamide POSS® shows a higher interaction with PVA, inducing lower values in the CO$_2$ flux. We have concluded that the interaction of the POSS® nanoparticles increased the crystallinity of the composite membranes, thereby playing an important role in the gas separation performance. Moreover, these nanocomposite membranes did not show separation according to a facilitated transport mechanism as expected, based on their functionalized amino-groups, thus, solution-diffusion was the main mechanism responsible for the transport phenomena.

Keywords: POSS®; nanocomposite membranes; CO$_2$ separation; PVA

1. Introduction

Global warming is one of the major problems that face humanity. Global warming is caused by the emission of greenhouse gases, of which the main component is carbon dioxide (CO$_2$). CO$_2$ emissions have dramatically increased in the last 50 years, and are still continually increasing each year [1]. The burning of fossil fuels contributes to most of the CO$_2$ emissions, and hence, there is a significant interest in developing technologies to reduce CO$_2$ emissions. Carbon capture and storage (CCS) has become an important instrument to reach the goals agreed on the 2015 Paris international convention [2]. Membrane technologies have seen an important growth in this market [3–6]. However, its application in large CO$_2$ capture processes in the power and industrial sector has not yet matured. Key challenges are related to the low partial pressure of CO$_2$ and the large scale required for flue gas treatment. For membranes to be cost-effective, further innovations in process design and membrane materials are thus needed. In this respect, the use of hybrid composite membranes that benefit from both an organic and inorganic part to further improve the gas separation performance have been widely studied. Membrane-based separation processes are not only cost effective and environmentally friendly, but also offer much more versatility and simplicity for customizing system designs with many novel polymeric materials now available. Moreover, membrane modularity and easy scale-up opens for retrofitting existing plants, as well as flexibility with respect to the CO_2 capture rate. The technology has, as well, a near instantaneous response, and a high turndown is possible that preserves plant operability.

The ability to selectively remove one component in a mixture while rejecting others describes the perfect separation device [7]. Various materials and methods are being developed for capture and storage, and in some cases, conversion, to mitigate the effect of global warming. Emerging and already established concepts for CO_2 capture encounter the challenge of finding an economically feasible and efficient separation technology from effluent gas streams [8,9]. The study of advanced materials and modern manufacturing methods have helped to obtain new and improved membranes that have better separation performances, thus contributing to obtain environmental friendly gas membrane separation processes which demand less energy. In this way, due to the low cost and easy processability of polymeric membranes, they have, so far, been a promising alternative for the non-condensable gas separation. Today, only a few polymer membrane materials are being used in industrial applications. However, in research, various polymers are being studied and reported to be potential candidates for use on commercial scale, due to good results achieved on lab scale [10]. Membranes require a high gas permeation rate combined with a high selectivity for process applications. Incrementing the performance of the membrane by increasing the permeance, will result in requiring less membrane area, which will help to reduce the costs and environmental imprint. Variables, such as mechanical and thermal stability of the material at the operating conditions, should also be considered when selecting the membrane material. The proper selection of a highly stable material, with high performance, has made the effort complex, and the research for an optimum material is continuously in progress [11–15].

The recognized trade-off between gas permeability and selectivity has resulted in a rather slow development of new polymeric membranes [16]. Nevertheless, both parameters need to be considered to have a promising membrane for industrial processes. In this case, the use of novel hybrid materials for making membranes that incorporate both an organic and an inorganic part, have opened new opportunities to develop new materials, and hence, bypass the limits that exist today. Both, the organic part and the inorganic part will, in different ways, contribute to the transport of gases through the hybrid membrane, while additionally, the inorganic part may contribute to mechanical and thermal strength. A lot of research has been published on the addition of zeolites to polymers for gas separation as mixed matrix membranes (MMMs) [17–19]. For instance, ZIF-71 nanoparticles, embedded into PEOT/PBT copolymer, have shown capability for dehumidification and CO_2 separation [20] in industrial use, and polymers of intrinsic microporosity (PIMs) are used to obtain membranes with high gas permeability [21].

On the other hand, facilitated transport membranes (FTMs) use a carrier to increase the transport of the desired gas through a membrane, without having a negative effect on the selectivity. Many researchers have studied this type of membrane [5,22–25].

In this study, facilitated transport was expected for CO_2 based on the selected nanoparticles. The manufactured membranes involve the addition of carriers that are expected to improve the gas separation properties. Nanosized particles, such as polyhedral oligomeric silsesquioxanes (POSS®) embedded in a polymer matrix, are reported here. POSS® nanoparticles have a rigid cage-like structure which is an intermediate between silica and siloxane. POSS® materials are well-defined, three-dimensional nano building blocks that can contribute to create unique hybrid materials, with a precise control of nanostructure and properties [26]. POSS® chemical reagents are nanostructured, with sizes of 1–3 nm. However, unlike silica or silicones, POSS® can be produced with nonreactive substituents, to make them more or less compatible with different monomers or polymers, depending on the way they are obtained [27]. The amino and amidino groups of the chosen POSS® were expected to provide the suitable carriers for CO_2 when humidity is present. This feature makes POSS® an attractive candidate for the fabrication of hybrid materials. POSS® has been used in several gas separation studies in recent years [28–31].

In our previous work [32], we have tested amino POSS® without achieving any improvement in CO_2 separation. In our effort to increase the performance of the membranes, we have modified these amino nanoparticles into the amidino and lactamide particles reported here. The aim of the current work reported here is to investigate the effect on the gas separation performance of the membranes by the addition of two different functionalized POSS® nanoparticles; one with amidine groups (Amidino POSS®) and the other with amino and lactamide functional groups (Lactamide POSS®). Amino groups are versatile and can undergo different chemical reactions. The functionalization of POSS® can thus be tailor-made, depending on the aimed properties. Amidino POSS® has been specially developed to enhance CO_2 capture capacity. Amidino POSS® and Lactamide POSS® would be expected to exhibit different mechanisms toward CO_2 capture. While Lactamide POSS® would catch CO_2 through forming a competition between carbonate and carbamate, Amidino POSS® would only form carbonate which is beneficial in term of cyclic capacity [32]. Moreover, functionalization of aminopropyl POSS® with lactic moieties is believed to improve compatibility of Lactamide POSS® with PVA, as it contains secondary hydroxyl groups.

2. Materials and Methods

2.1. Materials

Polyvinyl alcohol (PVA) was provided by Sigma-Aldrich, the molecular weight of 85,000–124,000, 87–89% hydrolyzed. Polysulfone (PSf) porous support flat sheet membranes with a molecular weight cut-off 50,000 Da were purchased from Alfa Laval (Denmark). The PSf support was washed. Two different types of nanoparticles belonging to the class of the polyhedral oligomeric silsesquioxane (POSS®) were synthesized. Figure 1 shows the synthesis of the two different types and their functionalization. The aminopropyl POSS® was synthesized by classical sol-gel synthesis from the 3-(aminopropyl)-triethoxysilane. Aminopropyl POSS® particles were subsequently functionalized in solvent-free reactions with *N,N*-dimethylacetamide dimethyl acetal to obtain Amidino POSS® and with ethyl lactate to prepare Lactamide POSS®. Amidino POSS® and Lactamide POSS® are obtained in *n*-propanol with a solid content of 42% for Amidino POSS® and 32% for Lactamide POSS®.

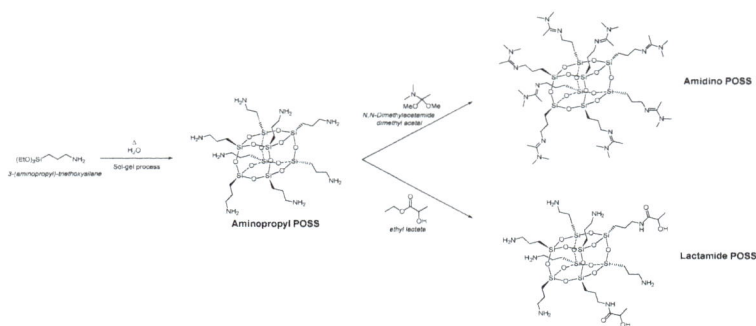

Figure 1. Synthesis of Amidino POSS® and Lactamide POSS®, respectively.

2.2. Membrane Preparation

PVA (3.0 g) was added to distilled water (97 g). The mixture was refluxed for two hours to dissolve the polymer under stirring at 500 rpm at 90 °C. Then, the PVA solution was cooled down to room temperature. Amidino POSS® and Lactamide POSS® solutions were stabilized separately, by stirring at 500 rpm for at least 15 min, before being mixed with the PVA solution. These nanoparticle solutions were then added dropwise to the PVA solution by filtering through a PTFE 5 μm syringe filter according to the required POSS®/PVA ratio. Afterwards, the solution was filtered again through a 5 μm PTFE syringe filter, and subsequently stirred for at least 30 min. Five centimeter diameter membranes of PSf were cut. The flat sheet composite membranes were prepared by casting 1 mL of the solution over the PSf disc. They were air dried overnight at 45 °C, and further air dried at 100 °C for one hour before a natural cooling, until the oven had reached 40 °C. These flat sheet composite membranes were used for SEM morphology characterization and gas permeation performance evaluation. The self-supported membrane samples for DSC, DMA, and FTIR characterization were prepared following the same temperature profile by using 5 mL of the POSS®/PVA solution in a glass petri dish.

2.3. Membrane Characterization

2.3.1. Scanning Electron Microscopy (SEM)

The thickness and surface of the composite membranes were studied using a Hitachi TM3030 and a Nova NanoSEM650 (FEI corp., Tokyo, Japan) field emission gun scanning electron microscope (FEG-SEM, Tokyo, Japan). The cross-section of the samples was prepared by freeze-fracturing/cutting in liquid nitrogen. The samples were sputter coated with a thin gold layer (90 s) to provide electronic conductivity to the samples.

2.3.2. Differential Scanning Calorimetry (DSC)

The thermal properties of the membrane materials were investigated using a differential scanning calorimeter (DSC 214, Netzsch, Selb, Germany). A sample of about 10–12 mg was placed in an aluminum pan covered with a proper lid, together with a standard empty pan, into the DSC sample holder. To eliminate the thermal history and remove any residual solvent, the samples were first heated from room temperature to 110 °C at 10 K/min, kept at this temperature for 10 min, and cooled to −50 °C at a rate of 10 K/min. The samples were then heated up from room temperature to 300 °C at a heating rate of 10 K/min, under an N_2 atmosphere. The analysis was then carried out from the second heating scan.

2.3.3. Thermogravimetric Analysis (TGA)

The thermal stability of the membranes was investigated with a thermogravimetric analyzer (TG209 F1, Netzsch, Selb, Germany). Around 10–12 mg of sample was placed in the sample pan and heated from 30 °C to 800 °C (10 K/min). Nitrogen was used as both the balance and sweep gas, with flow rates of 20 mL/min.

2.3.4. Fourier Transform Infrared Spectroscopy (FTIR)

FTIR was used to investigate the possible interaction between POSS® nanoparticles with the composite membranes. The IR spectroscopy experiments were carried out with a FTIR spectrometer (iS50 FT-IR, Nicolet, Thermo Fisher Scientific, Waltham, MA, USA) with a smart endurance reflection cell.

2.3.5. Dynamic Mechanical Analysis (DMA)

Dynamic mechanical analysis was conducted on a DMA 242 E Artemis (Netzsch, Selb, Germany) in tension mode under the following conditions: frequency 1 Hz, a proportional factor of 1.1, absolute target amplitude of 40 μm, a maximum dynamic force of 2.182 N, and no additional static force. To eliminate the thermal history, the samples were first heated from room temperature to 80 °C at 5 K/min, kept at this temperature for 10 min, and then cooled afterwards to −50 °C at a rate of 2 K/min, to be heated up again to 350 °C at a rate of 2 K/min. The samples were cut with 10 mm length and 5 mm width, with a thickness between 0.020–0.060 mm. With DMA, various transition temperatures can be determined while the mechanical properties are measured under an oscillatory strain, which supplies information about major transitions, as well as secondary and tertiary transitions not identifiable by other methods [33].

2.3.6. Gas Permeation Performance Evaluation

The membrane performance was measured in a feed gas mixture at a pressure of 1.3, 2, and 3 bars with a mixed gas permeation test rig, described elsewhere [34]. The experimental specifications, such as flow set parameters, temperature, and gas composition in feed and sweep streams, can be found in our previous work [32]. The CO_2 flux and CO_2/N_2 selectivity were calculated from the measured permeate CO_2 and N_2 concentration in the sweep flow gas.

The mixed gas permeability of CO_2 and N_2 was calculated assuming perfect mixing through Equation (1).

$$P_A = \frac{J_A \cdot \ell}{x_{fA} \cdot P_f - x_{pA} \cdot P_p} \tag{1}$$

where P_A represents the permeability (Barrer) of component A (CO_2 or N_2), J_A is the flux ($m^3(STP)/m^2 \cdot h$), x_{fA} and x_{pA}, represent molar fraction of the component on feed and permeate side respectively, P_f and P_p (bar) are the absolute pressure on feed and permeate side, and ℓ (μm) is the selective measured layer thickness.

The CO_2/N_2 selectivity $\alpha_{A/B}$, which corresponds to the ratio of the permeabilities for gases A (CO_2) and B (N_2), is given as follows:

$$\alpha_{\frac{A}{B}} = \frac{P_A}{P_B}. \tag{2}$$

3. Result and Discussion

3.1. Morphology of the Membranes

3.1.1. SEM

The composite dense layer deposited on the porous support can be seen in Figures 2 and 3, where examples are given for PVA, Amidino POSS®/PVA: 5%, and Lactamide POSS®/PVA: 5%. The dense composite membranes have an average thickness of 1–4 μm. The correct thickness of each membrane has been taken into account when the permeability is calculated. After scanning several membranes and places of the membrane, we conclude that a continuous composite PVA layer is formed on the support, and no evidence of pores or defects was observed. The white points in Figure 3A–C are most likely dust coming from the preparation of the SEM sample.

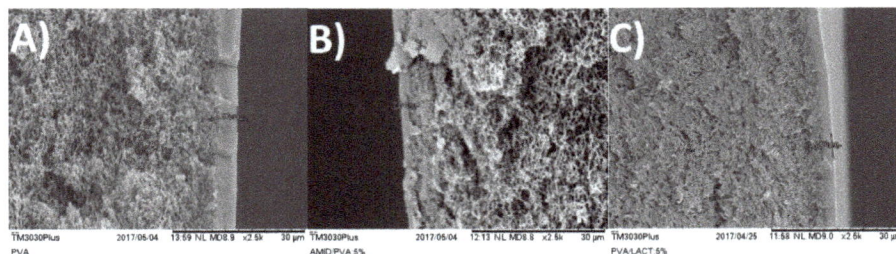

Figure 2. SEM-SE cross section pictures from the PSf support flat sheet membranes (**A**) PVA; (**B**) Amidino POSS®/PVA:5%; (**C**) Lactamide POSS®/PVA: 5%.

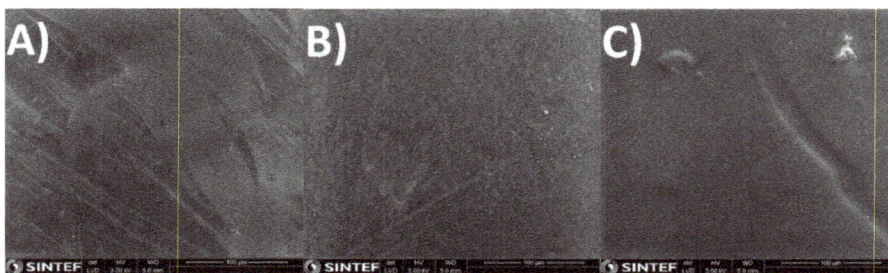

Figure 3. SEM-SE surface pictures from the PSf support flat sheet membranes (**A**) PVA; (**B**) Amidino POSS®/PVA: 5%, Lactamide POSS®/PVA: (**C**) 5%.

3.1.2. FTIR

Figures 4 and 5 show the FTIR spectra for the Amidino POSS®/PVA and Lactamide POSS®/PVA systems, respectively. The main IR peaks are given in Table 1. The characteristic peaks of PVA are seen at 1142, 1720, 2850, and 3260 cm^{-1}, the peaks for residual acetyl groups 1740, 1265 cm^{-1} decrease with addition of nanoparticles, since ratio PVA/nanoparticle loading is increasing, and thus, less PVA is present, and the residual acetyl peak decreases. The other peaks can be ascribed to either Amidino or Lactamide POSS® functionalities. The peak in the range of 1080 cm^{-1} corresponds to the asymmetric (Si–O–Si) stretching vibration band, belonging to the silica cage of POSS®. Its intensity increases with the nanoparticle loading [35].

Table 1. Main infrared peaks of the nanocomposite membranes [36–42].

Frequency (cm^{-1})	Bond Type
3260	O–H stretching (PVA)
2910–2942	CH$_2$ asymmetric/symmetric stretching (PVA)
1740, 1265	Residual acetyl group
1615	NH$_2$ scissoring
1656	N–H bending from amidine–lactamide
1400	N=N bending (amidine)
1142	PVA crystallites
1100	C–O stretching
1080	Si–O–Si asymmetric stretch
1020	Aliphatic C–N stretching (lactamide)

The absorption peak at approximately 1142 cm^{-1} corresponds to the PVA crystallites [36], increasing with the nanoparticle concentration. This is due to the interaction of the nanoparticles that work as a nucleating agent, which is more deeply explained in Section 3.2.2. The range at about 1400 cm^{-1} belongs to the azo compound N=N stretching of the Amidino POSS® nanoparticles, and shows a visible increment for the spectra of 25 and 50% ratio of nanoparticles. Finally, the range between 1550–1650 cm^{-1} belongs to the secondary amides bending, that also increases with the addition of nanoparticles. The Lactamide POSS®/PVA system shows the aliphatic stretch for C–N at 1020 cm^{-1}, the presence of POSS® at 1080 cm^{-1}, and the secondary amines bending, that increases with the addition of nanoparticles between 1550–1650 cm^{-1}. These spectra confirm the interaction between POSS® nanoparticles with the polymer in the composite membranes, which is influenced by the loading concentration of nanoparticles.

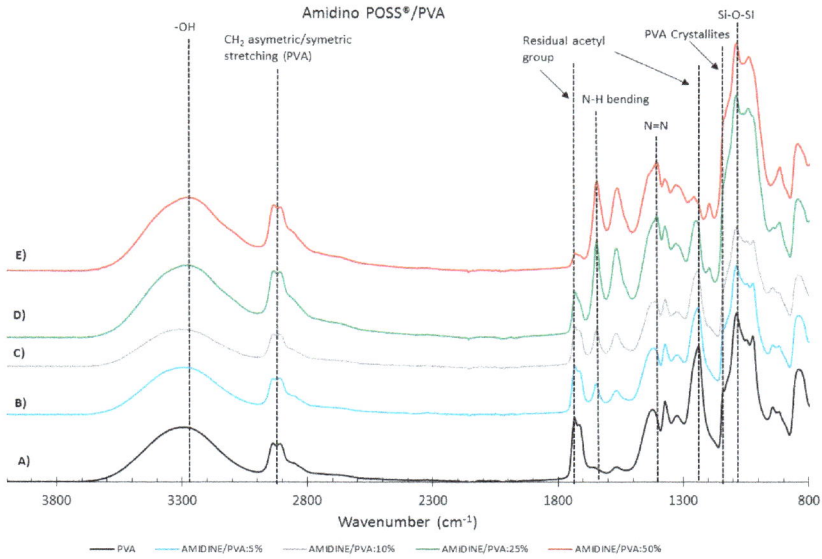

Figure 4. FTIR reflectance of (A) pure PVA, Amidino POSS®/PVA: (B) 5%; (C) 10%; (D) 25%; (E) 50%. The curves are shifted vertically for clarity.

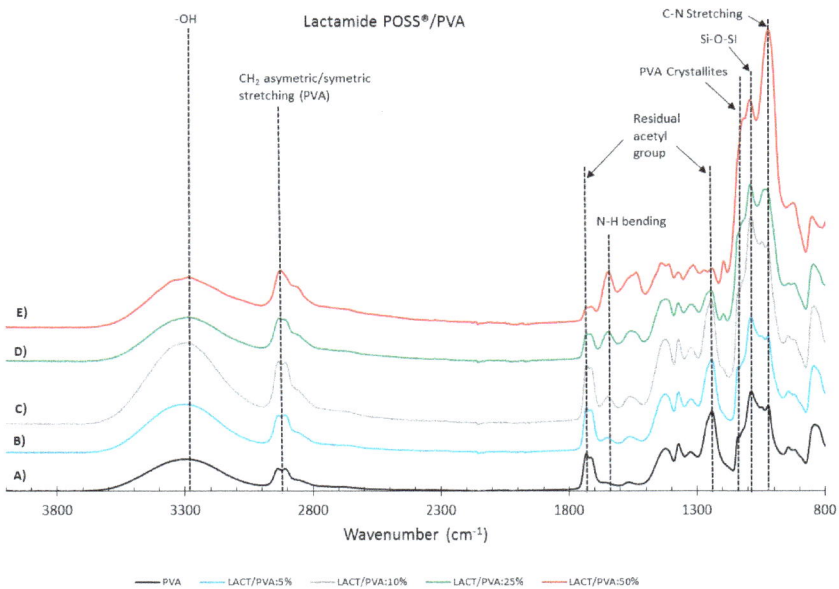

Figure 5. FTIR reflectance of (A) pure PVA, Lactamide POSS®/PVA: (B) 5%; (C)10%; (D) 25%; (E) 50%. The curves are shifted vertically for clarity.

3.2. *Thermomechanical Properties*

3.2.1. Thermogravimetric Analysis (TGA)

The thermal stability of the nanocomposite membranes with various nanoparticle loadings was studied by comparing the decomposition onset temperatures presented in Figures 6 and 7 for the Amidino POSS®/PVA and Lactamide POSS®/PVA systems, respectively. The pure PVA sample shows a major weight loss around 280 °C, attributed to the start of decomposition of the main polymer chain ending around 500 °C. The Amidino POSS®/PVA system shows two different major weight losses. The first one lies between 220 and 250 °C, attributed to the decomposition of the NH_2 groups and the start of the decomposition of the polymer chain, continuing with a second step at 400–450 °C, due to the byproducts generated by PVA during degradation [43,44]. The addition of the nanoparticles decreases this first step, however, the second weight loss, between 400–410 °C, increases with the addition of the nanoparticles. These weight losses correspond to the degradation of the polymer main chain, and the generated byproducts of the PVA. The systems end with different residuals for every sample, increasing with the nanoparticle loading. On the other hand, membranes prepared with Lactamide POSS® nanoparticles present a similar thermal stability as two major weight losses are observed in the range between 400–420 °C. The lactamide groups increase, slightly, the degradation of the polymer chain in this step, to reach a different final residual as well, depending on the concentration of the loading. The number corresponds initially to the residual of the PVA, and then to the increase of residual silicon content by the addition of nanoparticles. The weight loss between 100–200 °C in the samples containing nanoparticles could be due to either condensation of open POSS structures generating water, or due to the release of CO_2 already reacted with the amine and amidine groups from Amidino POSS® and Lactamide POSS® [45]. These decreased thermal stabilities can be explained by the increase in mobility of the PVA chains in the nanocomposite membranes. The chain transfer reaction will then be promoted, and consequently, the degradation process will be faster with the decomposition taking place at lower temperatures. However, the final complete degradation takes place at higher temperatures by the addition of the nanoparticles, which also change depending on the functionalization, corresponding to the Lactamide POSS® having the highest final degradation temperature.

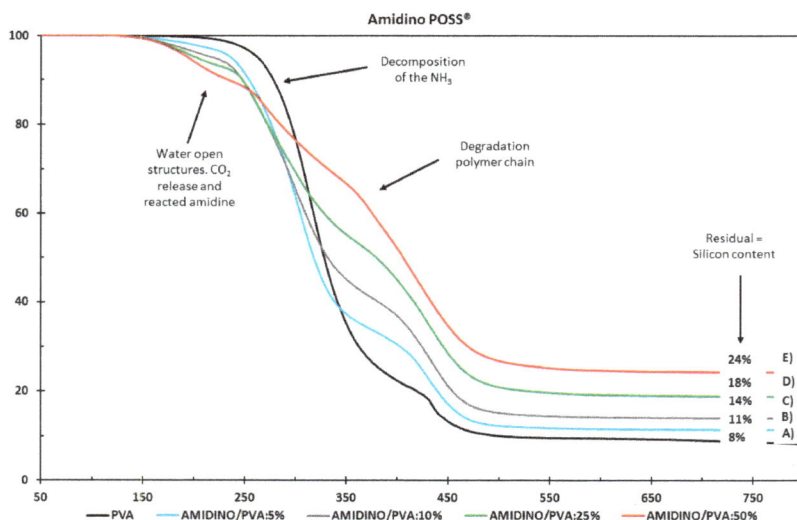

Figure 6. Thermogravimetric analysis of (A) PVA, Amidino POSS®/PVA: (B) 5%; (C) 10%; (D) 25%; (E) 50%.

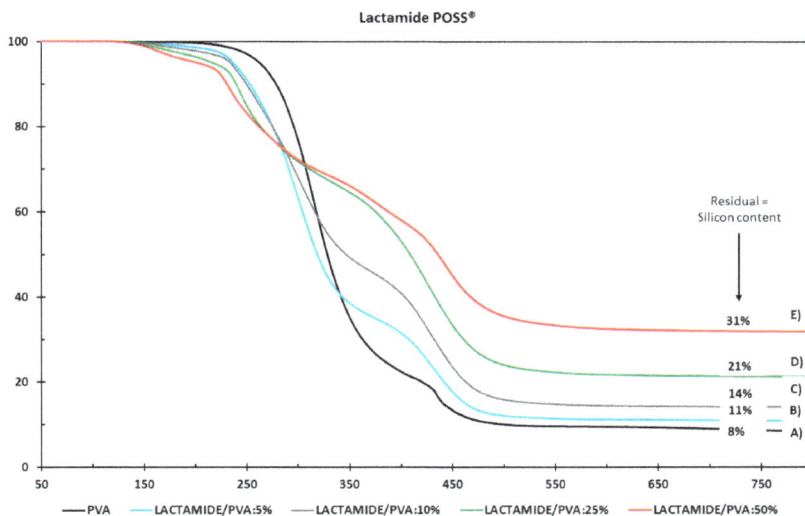

Figure 7. Thermogravimetric analysis of (A) PVA, Lactamide POSS®/PVA: (B) 5%; (C) 10%; (D) 25%; (E) 50%.

3.2.2. Differential Scanning Calorimetry (DSC)

Figure 8 shows the DSC thermograms for the melting enthalpies for the Amidino POSS®/PVA and Lactamide POSS®/PVA. The degree of crystallinity (X_c) is calculated from the ratio $X_c = \left(\Delta H_f / \Delta H_{f°} \right)$, where ΔH_f is the melting enthalpy measured, and $\Delta H_{f°}$ the 100% crystalline melting enthalpy, respectively. The melting enthalpy ($\Delta H_{f°}$) of a 100% crystalline PVA, is taken as 138.6 J/g [39].

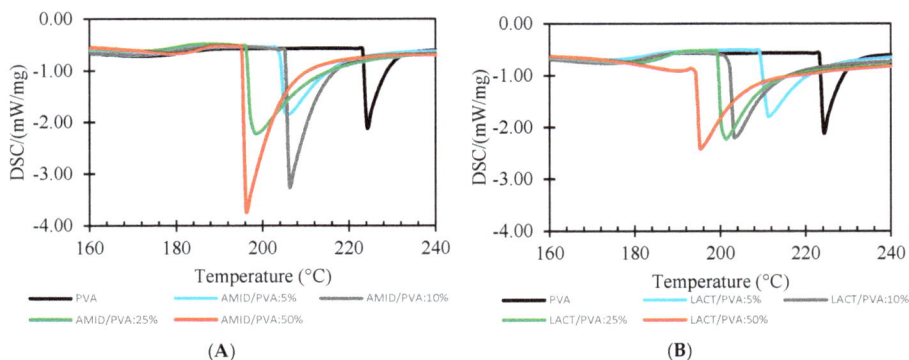

Figure 8. DSC analysis of Amidino POSS®/PVA (**A**); Lactamide POSS®/PVA (**B**). PVA (black), 5% (blue), 10% (gray), 25% (green), 50% (red).

Table 2 lists the values for the glass transition temperatures T_g, enthalpy and melting temperature, and degree of crystallinity (X_c) of the different nanocomposite membranes for the system Amidino POSS®/PVA and Lactamide POSS®/PVA, respectively. The glass transition increases for a loading of 5%, due to the rigidification of the polymer chain by the addition of the nanoparticles. However, when reaching 10%, the glass transition starts to decrease, due to the increment on the degree of the crystallinity. When reaching 25%, the T_g increases further, and at this point, the disturbance of the crystalline regions is higher as more crystalline segments are found—this is detectable by the increment in the degree of crystallinity. The interaction of the nanoparticles with the polymer is visible by the way the polymer chains packed, changing the type of lamella formation, and by the decrease in the melting temperature (T_m). The melting point (T_m) of the composite membranes is lower than the virgin PVA polymer membrane. Thus, the melting point of the composite material decreases because of the addition of the POSS® nanoparticles. The decrease in T_m with the addition of POSS® indicates that incorporation of POSS® encourages the crystallization process, and results in crystallites with lower thermal stability. This is explained as POSS® acting as a nucleating agent, increasing the degree of crystallinity [31,46]. The obtained numerical values are higher than 100%, because the degree of crystallinity is calculated, dividing the melting enthalpy measured by the 100% crystalline melting enthalpy. More energy is required to melt the crystalline segments formed by the addition of the nanoparticles, thus, the melting enthalpy measured by the addition of the nanoparticles is higher than the pure 100% crystalline polymer.

Table 2. Glass transition temperatures, enthalpy and melting temperature, and degree of crystallinity of the different nanocomposite membranes.

Sample	Loading (POSS®/PVA Ratio)	T_g (°C)	ΔH_f (J/g)	T_m (°C)	X_c(%)
PVA	0	70.5	51.7	223.4	37.3
Amidino POSS®	5	76.9	81.6	205.8	58.9
	10	63.2	106.5	205.7	76.8
	25	53.9	125.8	198.6	90.8
	50	67.0	151.6	195.5	109.4
Lactamide POSS®	5	77.8	74.4	211.3	53.7
	10	66.9	98.9	203.4	71.3
	25	55.7	134.0	201.0	96.7
	50	72.4	178.3	185.0	128.6

3.2.3. Dynamic Mechanical Analysis (DMA)

Storage Modulus (E′)

Table 3 shows the values of the storage modulus for the two types of nanocomposite membranes. The storage modulus (E′) is equivalent to Young's modulus of elasticity, which could be an indication of the hardness of the material. This is the response of the sample after receiving the strain, giving the amount of energy required to do so, expressed as modulus [33].

The composite membrane shows a typical thermoplastic behavior. The storage modulus is higher for the pure polymer, and it decreases slightly with the increase of nanoparticle loading, corresponding to the primary relaxation associated with the glass–rubber transition of amorphous PVA. Then, when the loading reaches 50%, the E′ modulus increases as compared to the virgin PVA. In this case, this increment can be explained by the crystallization occurring by the addition of POSS® acting as nucleating agent. The difference in the values between the Amidino and Lactamide POSS® can be due to the lactic moieties that are believed to improve compatibility of Lactamide POSS® with PVA, as it contains secondary hydroxyl groups.

Table 3. Storage modulus (E′) for the different nanocomposite membranes.

		PVA	% Loading							
			Amidino POSS®				Lactamide POSS®			
			5%	10%	25%	50%	5%	10%	25%	50%
Storage Modulus (E′)	Value (GPa)	8.2	6.7	8.0	6.9	8.3	8.0	7.3	7.4	10.1
	Onset Temperature (°C)	61	51	61	48	46	59	59	46	53

Tan Delta (Tand)

Figure 9 shows the change in tan delta by the change in the percentage of loading for the two types of nanoparticles, Amidino POSS®/PVA, and Lactamide POSS®/PVA, respectively.

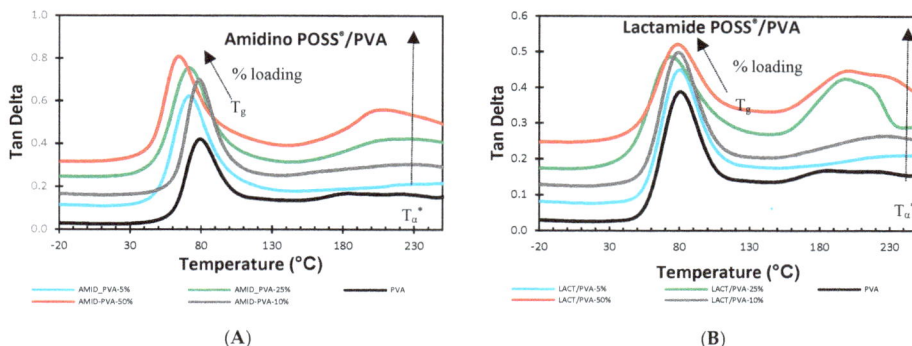

Figure 9. DMA analysis of Amidino POSS®/PVA (**A**); Lactamide POSS®/PVA (**B**). PVA (black), 5% (blue), 10% (gray), 25% (green), 50% (red). The curves are shifted vertically for clarity.

The addition of the Amidino POSS® and Lactamide POSS® nanoparticles show that the tan delta gradually decreases by the loading of the nanoparticles. A second peak is found in the region above the T_g. This peak is also known to be the alpha star transition (T_α^*). In semi-crystalline polymers, this transition is associated with the slippage between crystallites [33,47]. When increasing the loading of POSS® nanoparticles, these materials produce a different type of crystalline segment, thus inducing a modification of the tan delta (Tand T_g) values, and then, an increase of the alpha start transition is visible induced by the movement (slippage) of the crystalline segments. These changes are confirmed in the DSC measurements, where the melting temperature (T_m) is decreasing as an indication of forming different lamellar conformation. The alpha star transition also increased with the nanoparticles loading, see Table 4. A similar behavior is also observed for the storage modulus (E'), see Table 3. On the other hand, the Lactamide POSS®/PVA system shows that the addition of these nanoparticles has a higher effect of interaction since the Tand T_g and the alpha star transition (T_α^*) are higher in comparison with those obtained in the Amidino POSS®/PVA system. Moreover, these interactions are also visible in the degree of crystallinity (X_c) obtained by DSC, where these values are higher at higher Lactamide POSS® loadings. This improved compatibility was expected as the Lactamide groups can interact better with the PVA polymer chain. Similar behavior is observed for the storage modulus (E').

Table 4. T_α^* (alpha star transition) & Tand T_g, for the different nanocomposite membranes.

		PVA	Amidino POSS®				Lactamide POSS®			
			5%	10%	25%	50%	5%	10%	25%	50%
T_α^*	Absolute Value	0.17	0.14	0.16	0.20	0.26	0.16	0.17	0.23	0.27
Tand T_g	Absolute Value	0.42	0.54	0.56	0.54	0.50	0.40	0.40	0.34	0.30
	Onset Temperature(°C)	79.4	71.8	78.6	71.3	64.1	80.1	79.2	74.1	78.7

3.3. Gas Separation Performance of the Nanocomposite Membranes

3.3.1. Effect of Pressure

The effect of the feed pressure could have an important significance on the gas separation for facilitated transport membranes. The facilitated transport is based on a reversible reaction of CO_2,

H_2O, and the fixed carrier. As a general trend, with increasing feed pressure, the concentration of CO_2 dissolved into the membrane, increases. As the CO_2 concentration increases, some carriers may be saturated, and fewer sites are available to facilitate the CO_2 transport, resulting in a decrease of CO_2 permeance [48,49]. In other words, the CO_2 permeance may decrease due to a high CO_2 partial pressure on the feed side saturating the carriers [22,34]. To investigate this behavior of the membranes in the lower feed pressure region, tests were carried out by varying the feed pressure between 1.3 and 3 bars as shown in Figure 10. The performance of the membranes shows a slight decrease in CO_2 permeability, while the selectivity increases with the pressure, whereas a bigger change is visible in the loading for lactamide 25% between the pressure range 1.3 to 2 bar; this is possibly due to some inaccuracy during measurement. Unlike CO_2 permeance, N_2 permeance decreases with increasing feed pressure, which is consistent with the characteristic of "solution-diffusion" mechanism [50], resulting in the little increment in selectivity. The virgin PVA membrane shows the best performance in term of both permeability and selectivity, which is in accordance with our previous research [32]. This trend clearly shows that the membranes do not follow a facilitated transport mechanism, instead, the gas transport is dominated by the solution-diffusion mechanism, and the nanoparticle loading is not contributing with additional CO_2 transport to the gas separation in the tested pressure range.

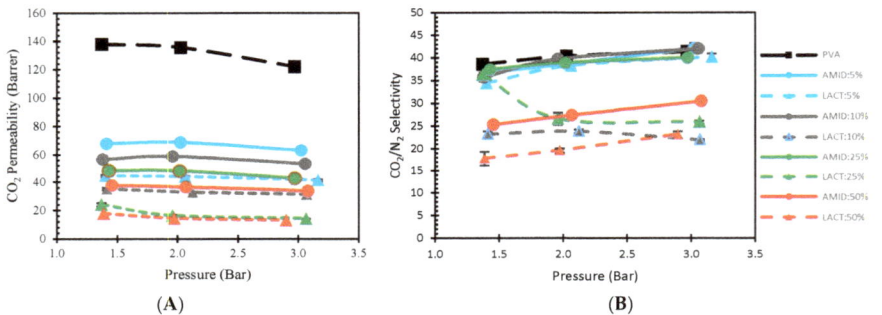

Figure 10. Effect of gas separation performance of nanocomposite membranes by the increment of pressure. Permeability (**A**); Selectivity (**B**). Amidino POSS®/PVA (straight line), Lactamide POSS®/PVA (dash line), PVA (black), 5% (blue), 10% (gray), 25% (green), 50% (red).

3.3.2. Effect of Nanoparticles Structure

By changing the length and the chemical structure of the substituents, POSS® can give different properties to the polymer, like thermal and chemical stability, and acts as a nucleating agent [51–54]. Figure 11 presents the CO_2 permeability as function of the degree of crystallinity (X_c), induced by the addition of the POSS® nanoparticles. The CO_2 permeability clearly decreases as the degree of crystallinity increases. Moreover, the permeability is always lower when Lactamide POSS® is added to the system. In other words, the higher the degree of crystallinity obtained with the lactamide groups, the lower the gas separation performance. The formation of crystalline segments makes the composite membrane less permeable. It is visible that the structure of the nanoparticle affects the permeance. Both membranes showed lower performance compared to the pure polymer. It seems that the addition of the nanoparticles hindered the CO_2 gas transport through the membrane. The most significant effect was observed at high Lactamide POSS® loading of nanoparticles. At high loading, more crystalline segments are obtained, as they are gas impermeable, reducing the performance of the membrane.

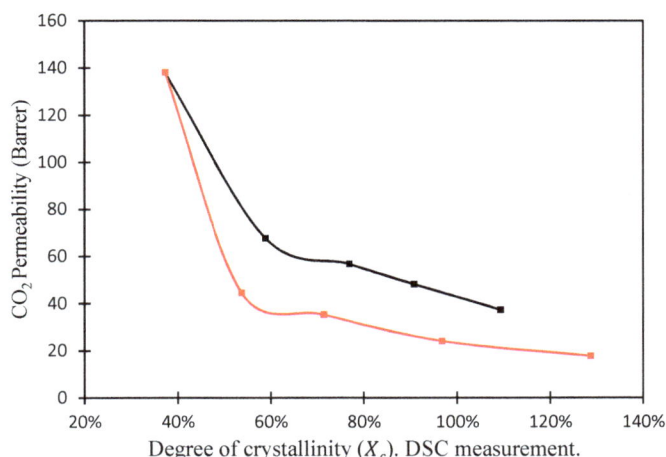

Figure 11. CO_2 permeability (1.3 bar) of nanocomposite membranes versus degree of crystallinity (X_c) DSC measurement. Amidino POSS®/PVA (black), Lactamide POSS®/PVA (red).

3.3.3. Effect of Loading Concentration

Figure 12 shows the CO_2 permeability and the degree of crystallinity (X_c) follow when increasing the nanoparticle loading. The crystalline regions on the PVA are impermeable to the penetrating molecules [55]. The addition of nanoparticles clearly shows an increment of the degree of crystallinity as the POSS® nanoparticles act as nucleating agent. The Lactamide POSS® group has a stronger interaction with the PVA, enhancing the chain packing order, increasing the degree of crystallinity. On the other hand, the CO_2 permeability has an opposite behavior, decreasing when increasing the nanoparticles loading. This decrease is reasonable, since these segments are less permeable. The addition of the nanoparticles does not contribute to any facilitated transport mechanism, moreover, it seems to obstruct the space the CO_2 can use to pass through the composite membrane, resulting in a decrease in CO_2 permeance. These changes in the gas separation properties could be also due to a decrease in the fractional free volume (FFV), and more research about this is needed.

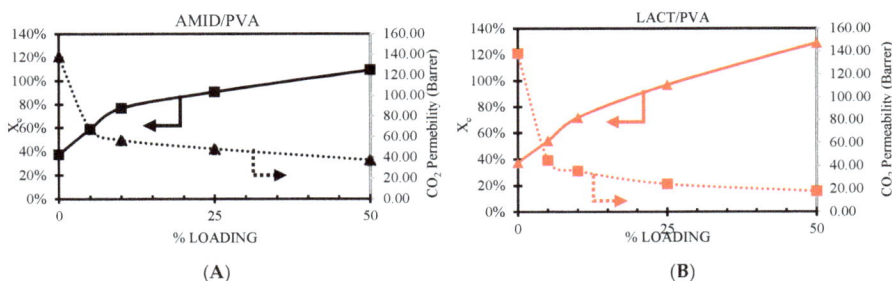

Figure 12. Effect of CO_2 permeability, dotted line, and degree of crystallinity (X_c), solid line, versus nanoparticle loading. Amidino POSS®/PVA (A), Lactamide POSS®/PVA (B).

4. Conclusions

Two different types of POSS® nanoparticles (Amidino POSS® and Lactamide POSS®) were used to understand the effect of the functional groups on the interaction with the polymer, and the gas separation performance. SEM pictures showed that a continuous composite PVA layer was formed on the support, and no evidence for pores or defects was observed. The FTIR spectra confirmed the structure of the resultant membranes, by showing the change by the addition of the nanoparticles. DSC studies indicated that the glass transition (T_g) decreased by the addition of the nanoparticle loading. It indicated an increase of the enthalpy of melting within the increase of the degree of crystallinity (X_c). The melting temperatures showed a decreasing trend, indicating a different lamellar conformation in the polymer chain. The change in T_g was also confirmed by the Tand T_g by DMA analysis. The addition of the nanoparticles has an effect on the storage modulus, however, it was not possible to observe any clear trend. A second peak $T_\alpha{}^*$ (alpha star transition) was found using DMA technique. This alpha transition is known to be the transition where the crystallites slip past one another. The value of this alpha transition increased when increasing the loading. In this sense, the Lactamide POSS® system showed a higher interaction with the polymer, as expected, by having a higher alpha star transition $T_\alpha{}^*$. The gas separation performance showed a decrease for the CO_2 permeability, while the selectivity slightly increased with the pressure. This trend shows that the membranes did not follow a facilitated transport mechanism, and the nanoparticle loading did not induce extra CO_2 transport to the gas separation. The system is primarily dominated by the solution-diffusion mechanism. The degree of crystallinity indicated a strong effect in the gas separation performance of the composite membranes. Finally, the permeability was lower when the Lactamide POSS® group was added to the system. This is due to the higher interaction with the polymer, which increases the degree of crystallinity and reduces the composite membrane permeability.

Author Contributions: Investigation, Writing-Original Draft Preparation, Writing-Review & Editing, G.G.; Writing-Review & Editing, Supervision, M.-B.H.; Writing-Review & Editing, Supervision, C.S.; Writing-Review & Editing, T.P.; Writing-Review & Editing, N.R. and Writing-Review & Editing, C.D.

Funding: The support from the Research Council of Norway (RCN) through the CLIMIT program (Project No: 224934) and BIGCCS Centre for Environment-friendly Energy Research (Project No: 193816/S60).

Acknowledgments: Martin F. Sunding (SINTEF Industry) is greatly acknowledged for the assistance with the SEM pictures.

Conflicts of Interest: The authors declare no conflicts of interest.

References

1. Powell, C.E.; Qiao, G.G. Polymeric CO_2/N_2 gas separation membranes for the capture of carbon dioxide from power plant flue gases. *J. Membr. Sci.* **2006**, *279*, 1–49. [CrossRef]

2. Rogelj, J.; den Elzen, M.; Höhne, N.; Fransen, T.; Fekete, H.; Winkler, H.; Schaeffer, R.; Sha, F.; Riahi, K.; Meinshausen, M. Paris agreement climate proposals need a boost to keep warming well below 2 °C. *Nature* **2016**, *534*, 631–639. [CrossRef] [PubMed]

3. Hägg, M.-B.; Lindbråthen, A. CO_2 capture from natural gas fired power plants by using membrane technology. *Ind. Eng. Chem. Res.* **2005**, *44*, 7668–7675. [CrossRef]

4. Haider, S.; Lindbråthen, A.; Hägg, M.-B. Techno-economical evaluation of membrane based biogas upgrading system: A comparison between polymeric membrane and carbon membrane technology. *Green Energy Environ.* **2016**, *1*, 222–234. [CrossRef]

5. Hussain, A.; Hägg, M.-B. A feasibility study of CO_2 capture from flue gas by a facilitated transport membrane. *J. Membr. Sci.* **2010**, *359*, 140–148. [CrossRef]

6. White, L.S.; Wei, X.; Pande, S.; Wu, T.; Merkel, T.C. Extended flue gas trials with a membrane-based pilot plant at a one-ton-per-day carbon capture rate. *J. Membr. Sci.* **2015**, *496*, 48–57. [CrossRef]

7. Gnanasekaran, D.; Reddy, B.S.R. Cost effective poly(urethane-imide)-poss membranes for environmental and energy-related processes. *Clean Technol. Environ. Policy* **2013**, *15*, 383–389. [CrossRef]

8. D'Alessandro, D.M.; Smit, B.; Long, J.R. Carbon dioxide capture: Prospects for new materials. *Angew. Chem. Int. Ed.* **2010**, *49*, 6058–6082. [CrossRef] [PubMed]

9. Ashley, M.; Magiera, C.; Ramidi, P.; Blackburn, G.; Scott, T.G.; Gupta, R.; Wilson, K.; Ghosh, A.; Biswas, A. Nanomaterials and processes for carbon capture and conversion into useful by-products for a sustainable energy future. *Greenh. Gases Sci. Technol.* **2012**, *2*, 419–444. [CrossRef]

10. Baker, R.W. Future directions of membrane gas separation technology. *Ind. Eng. Chem. Res.* **2002**, *41*, 1393–1411. [CrossRef]

11. Freeman, B.D.; Pinnau, I. Polymeric materials for gas separations. In *Polymer Membranes for Gas and Vapor Separation*; American Chemical Society: Washington, DC, USA, 1999; Volume 733, pp. 1–27.

12. Koros, W.J. Gas separation membranes: Needs for combined materials science and processing approaches. *Macromol. Symp.* **2002**, *188*, 13–22. [CrossRef]

13. Gupta, Y.; Hellgardt, K.; Wakeman, R.J. Enhanced permeability of polyaniline based nano-membranes for gas separation. *J. Membr. Sci.* **2006**, *282*, 60–70. [CrossRef]

14. Bernardo, P.; Drioli, E.; Golemme, G. Membrane gas separation: A review/state of the art. *Ind. Eng. Chem. Res.* **2009**, *48*, 4638–4663. [CrossRef]

15. Trong Nguyen, Q.; Sublet, J.; Langevin, D.; Chappey, C.; Marais, S.; Valleton, J.-M.; Poncin-Epaillard, F. CO_2 permeation with pebax®-based membranes for global warming reduction. In *Membrane Gas Separation*; John Wiley & Sons, Ltd.: Hoboken, NJ, USA, 2010; pp. 255–277.

16. Robeson, L.M. The upper bound revisited. *J. Membr. Sci.* **2008**, *320*, 390–400. [CrossRef]

17. Sorribas, S.; Zornoza, B.; Téllez, C.; Coronas, J. Mixed matrix membranes comprising silica-(ZIF-8) core–shell spheres with ordered meso–microporosity for natural- and bio-gas upgrading. *J. Membr. Sci.* **2014**, *452*, 184–192. [CrossRef]

18. Thompson, J.A.; Vaughn, J.T.; Brunelli, N.A.; Koros, W.J.; Jones, C.W.; Nair, S. Mixed-linker zeolitic imidazolate framework mixed-matrix membranes for aggressive CO_2 separation from natural gas. *Microporous Mesoporous Mater.* **2014**, *192*, 43–51. [CrossRef]

19. Nafisi, V.; Hägg, M.-B. Development of dual layer of ZIF-8/pebax-2533 mixed matrix membrane for CO_2 capture. *J. Membr. Sci.* **2014**, *459*, 244–255. [CrossRef]

20. Yong, W.F.; Ho, Y.X.; Chung, T.S. Nanoparticles embedded in amphiphilic membranes for carbon dioxide separation and dehumidification. *ChemSusChem* **2017**, *10*, 4046–4055. [CrossRef] [PubMed]

21. Yong, W.F.; Li, F.Y.; Chung, T.S.; Tong, Y.W. Molecular interaction, gas transport properties and plasticization behavior of CPIM-1/torlon blend membranes. *J. Membr. Sci.* **2014**, *462*, 119–130. [CrossRef]

22. Ansaloni, L.; Zhao, Y.; Jung, B.T.; Ramasubramanian, K.; Baschetti, M.G.; Ho, W.S.W. Facilitated transport membranes containing amino-functionalized multi-walled carbon nanotubes for high-pressure CO_2 separations. *J. Membr. Sci.* **2015**, *490*, 18–28. [CrossRef]

23. Deng, L.; Hägg, M.-B. Swelling behavior and gas permeation performance of PVAm/PVA blend FSC membrane. *J. Membr. Sci.* **2010**, *363*, 295–301. [CrossRef]

24. Wu, H.; Li, X.; Li, Y.; Wang, S.; Guo, J.; Jiang, Z.; Wu, C.; Xin, Q.; Lu, X. Facilitated transport mixed matrix membranes incorporated with amine functionalized MCM-41 for enhanced gas separation properties. *J. Membr. Sci.* **2014**, *465*, 78–90. [CrossRef]

25. Washim Uddin, M.; Hägg, M.-B. Natural gas sweetening—The effect on CO_2–CH_4 separation after exposing a facilitated transport membrane to hydrogen sulfide and higher hydrocarbons. *J. Membr. Sci.* **2012**, *423–424*, 143–149. [CrossRef]

26. Markovic, E.; Constantopolous, K.; Matisons, J.G. Polyhedral oligomeric silsesquioxanes: From early and strategic development through to materials application. In *Applications of Polyhedral Oligomeric Silsesquioxanes*; Hartmann-Thompson, C., Ed.; Springer: Dordrecht, The Netherlands, 2011; pp. 1–46.

27. Pielichowski, K.; Njuguna, J.; Janowski, B.; Pielichowski, J. Polyhedral oligomeric silsesquioxanes (poss)-containing nanohybrid polymers. In *Supramolecular Polymers Polymeric Betains Oligomers*; Springer: Berlin/Heidelberg, Germany, 2006; pp. 225–296.

28. Dasgupta, B.; Sen, S.K.; Banerjee, S. Aminoethylaminopropylisobutyl poss—Polyimide nanocomposite membranes and their gas transport properties. *Mater. Sci. Eng. B* **2010**, *168*, 30–35. [CrossRef]

29. Li, F.; Li, Y.; Chung, T.-S.; Kawi, S. Facilitated transport by hybrid poss®–matrimid®–ZN^{2+} nanocomposite membranes for the separation of natural gas. *J. Membr. Sci.* **2010**, *356*, 14–21. [CrossRef]

30. Li, Y.; Chung, T.-S. Molecular-level mixed matrix membranes comprising pebax® and poss for hydrogen purification via preferential CO_2 removal. *Int. J. Hydrogen Energy* **2010**, *35*, 10560–10568. [CrossRef]

31. Rahman, M.M.; Filiz, V.; Shishatskiy, S.; Abetz, C.; Georgopanos, P.; Khan, M.M.; Neumann, S.; Abetz, V. Influence of poly(ethylene glycol) segment length on CO_2 permeation and stability of polyactive membranes and their nanocomposites with peg poss. *ACS Appl. Mater. Interfaces* **2015**, *7*, 12289–12298. [CrossRef] [PubMed]

32. Guerrero, G.; Hägg, M.-B.; Kignelman, G.; Simon, C.; Peters, T.; Rival, N.; Denonville, C. Investigation of amino and amidino functionalized polyhedral oligomeric silsesquioxanes (poss®) nanoparticles in pva-based hybrid membranes for CO_2/N_2 separation. *J. Membr. Sci.* **2017**, *544*, 161–173. [CrossRef]

33. Menard, K.P. *Dynamic Mechanical Analysis: A Practical Introductio*; CRC Press: Boca Raton, FL, USA, 2008.

34. Deng, L.; Kim, T.-J.; Hägg, M.-B. Facilitated transport of CO_2 in novel PVAm/PVA blend membrane. *J. Membr. Sci.* **2009**, *340*, 154–163. [CrossRef]

35. Ramírez, C.; Rico, M.; Torres, A.; Barral, L.; López, J.; Montero, B. Epoxy/poss organic–inorganic hybrids: Atr-ftir and dsc studies. *Eur. Polym. J.* **2008**, *44*, 3035–3045. [CrossRef]

36. Ben Hamouda, S.; Nguyen, Q.T.; Langevin, D.; Roudesli, S. Poly(vinylalcohol)/poly(ethyleneglycol)/poly(ethyleneimine) blend membranes—Structure and CO_2 facilitated transport. *C. R. Chim.* **2010**, *13*, 372–379. [CrossRef]

37. Mark, J.E. *Polymer Data Handbook*; Oxford University Press: Oxford, UK, 1999.

38. Yong, W.F.; Kwek, K.H.A.; Liao, K.-S.; Chung, T.-S. Suppression of aging and plasticization in highly permeable polymers. *Polymer* **2015**, *77*, 377–386. [CrossRef]

39. Yang, X.; Li, L.; Shang, S.; Tao, X.-M. Synthesis and characterization of layer-aligned poly(vinyl alcohol)/graphene nanocomposites. *Polymer* **2010**, *51*, 3431–3435. [CrossRef]

40. Chua, M.L.; Shao, L.; Low, B.T.; Xiao, Y.; Chung, T.-S. Polyetheramine–polyhedral oligomeric silsesquioxane organic–inorganic hybrid membranes for CO_2/H_2 and CO_2/N_2 separation. *J. Membr. Sci.* **2011**, *385*, 40–48. [CrossRef]

41. Fei, M.; Jin, B.; Wang, W.; Liu, L. Synthesis and characterization of ab block copolymers based on polyhedral oligomeric silsesquioxane. *J. Polym. Res.* **2010**, *17*, 19. [CrossRef]

42. Choi, J.; Yee, A.F.; Laine, R.M. Organic/inorganic hybrid composites from cubic silsesquioxanes. Epoxy resins of octa (dimethylsiloxyethylcyclohexylepoxide) silsesquioxane. *Macromolecules* **2003**, *36*, 5666–5682. [CrossRef]

43. Chen, C.-H.; Wang, F.-Y.; Mao, C.-F.; Liao, W.-T.; Hsieh, C.-D. Studies of chitosan: II. Preparation and characterization of chitosan/poly(vinyl alcohol)/gelatin ternary blend films. *Int. J. Biol. Macromol.* **2008**, *43*, 37–42. [CrossRef] [PubMed]

44. Holland, B.J.; Hay, J.N. The thermal degradation of poly(vinyl alcohol). *Polymer* **2001**, *42*, 6775–6783. [CrossRef]

45. Mondal, A.; Barooah, M.; Mandal, B. Effect of single and blended amine carriers on CO_2 separation from CO_2/N_2 mixtures using crosslinked thin-film poly(vinyl alcohol) composite membrane. *Int. J. Greenh. Gas Control* **2015**, *39*, 27–38. [CrossRef]

46. Heeley, E.; Hughes, D.; Taylor, P.; Bassindale, A. Crystallization and morphology development in polyethylene–octakis (n-octadecyldimethylsiloxy) octasilsesquioxane nanocomposite blends. *RSC Adv.* **2015**, *5*, 34709–34719. [CrossRef]

47. Hay, W.T.; Byars, J.A.; Fanta, G.F.; Selling, G.W. Rheological characterization of solutions and thin films made from amylose-hexadecylammonium chloride inclusion complexes and polyvinyl alcohol. *Carbohydr. Polym.* **2017**, *161*, 140–148. [CrossRef] [PubMed]

48. Zhao, J.; Wang, Z.; Wang, J.; Wang, S. Influence of heat-treatment on CO_2 separation performance of novel fixed carrier composite membranes prepared by interfacial polymerization. *J. Membr. Sci.* **2006**, *283*, 346–356. [CrossRef]

49. Kim, T.-J.; Li, B.; Hägg, M.-B. Novel fixed-site–carrier polyvinylamine membrane for carbon dioxide capture. *J. Membr. Sci. Part B Polym. Phys.* **2004**, *42*, 4326–4336. [CrossRef]

50. Wijmans, J.; Baker, R. The solution-diffusion model: A review. *J. Membr. Sci.* **1995**, *107*, 1–21. [CrossRef]

51. Raftopoulos, K.N.; Koutsoumpis, S.; Jancia, M.; Lewicki, J.P.; Kyriakos, K.; Mason, H.E.; Harley, S.J.; Hebda, E.; Papadakis, C.M.; Pielichowski, K.; et al. Reduced phase separation and slowing of dynamics in polyurethanes with three-dimensional poss-based cross-linking moieties. *Macromolecules* **2015**, *48*, 1429–1441. [CrossRef]

52. Ayandele, E.; Sarkar, B.; Alexandridis, P. Polyhedral oligomeric silsesquioxane (poss)-containing polymer nanocomposites. *Nanomaterials* **2012**, *2*, 445–475. [CrossRef] [PubMed]

53. Raftopoulos, K.N.; Janowski, B.; Apekis, L.; Pissis, P.; Pielichowski, K. Direct and indirect effects of poss on the molecular mobility of polyurethanes with varying segment mw. *Polymer* **2013**, *54*, 2745–2754. [CrossRef]

54. Raftopoulos, K.N.; Jancia, M.; Aravopoulou, D.; Hebda, E.; Pielichowski, K.; Pissis, P. Poss along the hard segments of polyurethane. Phase separation and molecular dynamics. *Macromolecules* **2013**, *46*, 7378–7386. [CrossRef]

55. Gholap, S.G.; Jog, J.P.; Badiger, M.V. Synthesis and characterization of hydrophobically modified poly(vinyl alcohol) hydrogel membrane. *Polymer* **2004**, *45*, 5863–5873. [CrossRef]

Article

Mixed Matrix Membranes of Boron Icosahedron and Polymers of Intrinsic Microporosity (PIM-1) for Gas Separation

Muntazim Munir Khan, Sergey Shishatskiy and Volkan Filiz *

Institute of Polymer Research, Helmholtz-Zentrum Geesthacht, Max-Planck-Strasse 1, 21502 Geesthacht, Germany; muntazim.khan@hzg.de (M.M.K.); sergey.shishatskiy@hzg.de (S.S.)
* Correspondence: volkan.filiz@hzg.de; Tel.: +49-41-5287-2425

Received: 21 November 2017; Accepted: 23 December 2017; Published: 2 January 2018

Abstract: This work reports on the preparation and gas transport performance of mixed matrix membranes (MMMs) based on the polymer of intrinsic microporosity (PIM-1) and potassium dodecahydrododecaborate ($K_2B_{12}H_{12}$) as inorganic particles (IPs). The effect of IP loading on the gas separation performance of these MMMs was investigated by varying the IP content (2.5, 5, 10 and 20 wt %) in a PIM-1 polymer matrix. The derived MMMs were characterized by scanning electron microscopy (SEM), thermogravimetric analysis (TGA), single gas permeation tests and sorption measurement. The $PIM1/K_2B_{12}H_{12}$ MMMs show good dispersion of the IPs (from 2.5 to 10 wt %) in the polymer matrix. The gas permeability of $PIM1/K_2B_{12}H_{12}$ MMMs increases as the loading of IPs increases (up to 10 wt %) without sacrificing permselectivity. The sorption isotherm in PIM-1 and $PIM1/K_2B_{12}H_{12}$ MMMs demonstrate typical dual-mode sorption behaviors for the gases CO_2 and CH_4.

Keywords: mixed matrix membranes; polymer of intrinsic microporosity; borane; gas separation membrane

1. Introduction

Membrane technology can potentially provide environmental and economic advantages to virtually any process dependent on gas separation. However, the ability to produce durable, large-area membranes at relatively low cost and the wider application of polymeric membranes is hindered by their intrinsic permeability and selectivity limitations. These limitations were first reported by Robeson as an upper bound trade-off between permeability and selectivity and later by Freeman [1,2]. Based on the need for a more efficient membrane than purely polymeric membranes, a new concept of mixed-matrix membranes (MMMs) has been proposed. MMMs are hybrid membranes containing solid, liquid, or both solid and liquid inorganic fillers embedded in a polymer matrix [3,4]. MMMs have the potential to achieve higher selectivity with equal or higher permeability compared to existing polymer membranes while maintaining their advantages of mechanical stability and the possibility of large-scale production. Compared to pure polymer membranes, many polymer-inorganic nanocomposite membranes containing silica, carbon nanotubes, zeolite, metal organic framework (MOF), titania, etc., as IPs show higher permeability without sacrificing gas selectivity [5,6]. However, there are still many issues that need to be addressed for the large-scale industrial production of MMMs. Attempts to enhance the compatibility between the inorganic and polymeric components by introducing mutually interactive functional groups to the polymer and the molecular sieve have led to partial blockage of the sieve pores, thus hindering separation performance.

Polynuclear boranes, another class of inorganic particles, have been extensively studied for the past fifty years and their chemistry is well-established and designated with the general formula

$B_nH_n{}^{2-}$ (where $n = 6$–12) [7,8]. Figure 1 shows an example of polynuclear boranes, i.e., the $[B_{12}H_{12}]^{2-}$ is a dianion and bicapped square antiprism *closo* structure and $B_{12}H_{12}{}^{2-}$ dianion has icosahedral *closo* geometry. Geometrically, polynuclear borane anions have trigonal faces. For example, icosahedral *closo*-$B_{12}H_{12}{}^{2-}$ consists of 12 boron atoms each bonded to five neighboring boron atoms within the icosahedron and to an external atom such as hydrogen. One or more BH vertices can be exchanged for isoelectronic CH^+ vertices, giving rise to a variety of carborane structures. Diverse functionalizations at the resulting CH vertices provide novel structures with unique applications in material science and biomedicine [9–12].

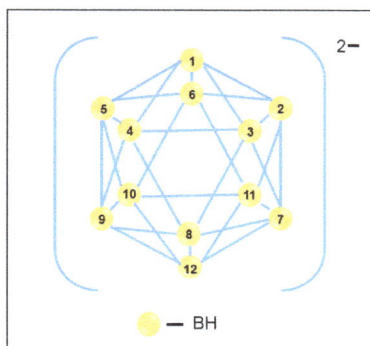

Figure 1. Polynuclear borane structure and numbering of atoms in the $[B_{12}H_{12}]^{2-}$ anion.

Tailoring free volume cavities by controlling the molecular weight and the structure of glassy polymers directly influences the gas transport properties [13]. In particular, a class of high free volume polymers were potential candidates for gas separation applications with the capability to optimize gas permeability and selectivity by changing the polymer chain packing. McKeown and Budd first reported a new class of rigid ladder-type polymers containing highly contorted chains and named them polymers of intrinsic microporosity (PIM) [14]. Among these materials, PIM-1 (Figure 2), containing the contorted angled spirobisindane unit and rigid polymer backbone and high free volume, which attracted the most attention due to the combination of outstanding permeability with relatively moderate but technically attractive permselectivity, especially for O_2/N_2 and CO_2/CH_4 pairs [15–17].

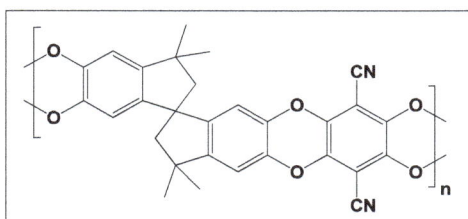

Figure 2. PIM-1 polymer structure.

In the present work, MMMs were fabricated by the incorporation of $K_2B_{12}H_{12}$ (as inorganic particles) into a PIM-1 matrix (as a polymer matrix). Pure gas permeability data (H_2, N_2, O_2, CO_2 and CH_4 gases) were reported for pristine PIM-1 and their MMMs. Physical properties such as the thermal analysis and morphology of the IPs were investigated. The prepared MMMs were characterized by scanning electron microscopy (SEM), thermogravimetric analysis (TGA), single gas permeation tests and gas sorption measurement. To the best of our knowledge, so far there is no MMM publication available on using boron icosahedron $B_nH_n{}^{2-}$ (as an IP) combined with PIM-1 (as a polymer matrix) for a gas separation membrane.

2. Theory and Background

2.1. Gas Sorption

In order to understand the gas transport properties of MMMs, two aspects need to be considered. First, static sorption experiments can reveal the maximum sorption capacity of a polymer for certain gas, which helps us to understand why IPs can enhance the performance of MMMs compared to pristine polymer membranes. Second, dynamic sorption experiments reveal information on the kinetics of the gas sorption from which diffusion coefficients can be determined.

2.1.1. Static Gas Sorption

Gas sorption in glassy polymeric membranes described by the dual-mode sorption model. In this model, penetrant molecules are viewed as being partitioned between the dense equilibrium structure of the polymer (dissolved mode) and the non-equilibrium excess volume of the glassy polymer (the so-called hole filling or Langmuir mode) [18]. The dual mode model is described by Equation (1):

$$C = C_D + C_H \tag{1}$$

where C is the total concentration of penetrant in the polymer (mol/g), C_D is the dissolved mode penetrant concentration, and C_H is the penetrant concentration in the hole filling of Langmuir mode. C_D is written as a linear function of pressure and C_H is expressed by a Langmuir isotherm to give:

$$C = k_D p + \left(\frac{C'_H b_P}{1 + b_P} \right) \tag{2}$$

where k_D is Henry's law/dissolved mode sorption constant [mol/(g·bar)], p the pressure (bar), C'_H is the Langmuir/hole filling capacity constant (mol/g) and b is the Langmuir affinity parameter (1/bar). The parameter k_D shows the penetrant dissolved in the polymer matrix at equilibrium and b characterizes the sorption affinity for a specific gas–polymer system. These parameters can be determined from the measured sorption data. C'_H is often used to measure the amount of non-equilibrium excess free volume in the glassy state [19].

2.1.2. Dynamic Gas Sorption

Diffusion coefficients can be accurately determined from the mass uptake curves (M_t/M_∞) by data-fitting Fick's second law for the sorption of penetrant in the film as described by Crank [20]:

$$\frac{M_t}{M_\infty} = 1 - \frac{8}{\pi} \sum_{n=0}^{\infty} \frac{1}{(2n+1)^2} exp \left[-\frac{D(2n+1)^2 \pi^2 t}{l^2} \right] \tag{3}$$

where M_t and M_∞ represent the amount of gas absorbed by the membrane film at time t and the equilibrium sorption after infinite time, respectively. D is the kinetic (transport) diffusion coefficient, t is the time required to attain M_t and l is the thickness of the sample.

2.2. Gas Permeation

Gas permeation through a dense membrane takes place according to the well-known solution–diffusion mechanism [21]:

$$P_i = S_i \times D_i \tag{4}$$

where the permeability coefficient (P_i) in Barrer (1Barrer = 10^{-10} cm^3(STP)·cm/(cm^2·s·cmHg)) is the product of the solubility coefficient (S_i) (cm^3(STP)/(cm^3·cmHg)) and the diffusion coefficient (D_i) (cm^2/s) of component i. The ideal selectivity for a gas pair is the ratio of their permeability coefficients:

$$\alpha_{ij} = \frac{P_i}{P_j} = \frac{S_i \times D_i}{S_j \times D_j} = \left(\frac{S_i}{S_j}\right) \times \left(\frac{D_i}{D_j}\right) \tag{5}$$

where D_i/D_j is the diffusion selectivity and S_i/S_j is the solubility selectivity of components i and j, respectively. Diffusion coefficients increase with a decrease in the penetrant size, increasing the polymer fractional free volume, increasing polymer chain flexibility, increasing the temperature and decreasing polymer–penetrant interactions [22]. On the other hand, solubility coefficients increase with increasing polymer–penetrant interactions, decreasing temperature and the increasing condensability of the penetrant.

3. Materials

The monomer 5,5′,6,6′-tetrahydroxy-3,3,3′,3″-tetramethyl-1,1′-spirobisindane (TTSBI, 97%) was supplied by ABCR, Karlsruhe, Germany and 2,3,5,6-tetrafluoroterephthalonitrile (TFTPN, 99%) was kindly donated by Lanxess (Cologne, Germany). TFTPN was sublimated twice under vacuum prior to use. Potassium carbonate (K$_2$CO$_3$ > 99.5%) was dried overnight under vacuum at 120 °C in order to ensure no moisture is trapped in it and then milled in a ball mill for 15 min. Potassium dodecahydrododecaborate hydrate (K$_2$B$_{12}$H$_{12}$·XH$_2$O > 98%) was obtained from Strem chemicals Inc. (Kehl, Germany) and bis-tetrabutylammonium *closo*-dodecahydrododecaborate [N(C$_4$H$_9$)$_4$]$_2$B$_{12}$H$_{12}$ was supplied by Technical University Darmstadt, Inorganic solid state chemistry department. Diethylbenzene (isomeric mixture) was purchased from Sigma-Aldrich (Steinheim, Germany), dimethylacetamide (DMAc > 99%), tetrahydrofuran (THF > 99.9%), methanol (MeOH > 99.9%), chloroform (CHCl$_3$ > 99.99%), dioxane (>99%), from Merck (Darmstadt, Germany) were used as received.

4. Experimental Section

4.1. Pristine PIM-1 Synthesis and Mixed Matrix Membranes Preparation

PIM-1 was synthesized by using the method described elsewhere [23–27]. PIM-1 and K$_2$B$_{12}$H$_{12}$ were dried in a vacuum oven at 120 °C overnight before use. The pristine PIM-1 membrane was prepared by mixing 2% (w/w) polymer in chloroform as a solvent. MMMs were prepared with K$_2$B$_{12}$H$_{12}$ with different weight ratios (2.5 wt %; 5 wt %; 10 wt %; 20 wt %) as determined by Equation (6).

$$IPs\ loading = \frac{wt.\ IP}{wt.\ IP + wt\ polymer} \times 100 \tag{6}$$

Considering a PIM-1 and K$_2$B$_{12}$H$_{12}$ density and assuming volumes are additive, the IPs volume fraction (ϕ_{IP}) were calculated according to Equation (7).

$$\phi_{IP} = \frac{\frac{w_{IP}}{\rho_{IP}}}{\frac{w_P}{\rho_P} + \frac{w_{IP}}{\rho_{IP}}} \tag{7}$$

where w_{IP} and w_P denote the weight of IPs and polymer, respectively, and ρ_{IP} and ρ_P are the density of IPs and polymer, respectively. For the MMMs fabrication, the K$_2$B$_{12}$H$_{12}$ was dispersed in chloroform by sonication using an ultrasonic bath (Bendelin, SONOREX Super, Bendelin Electronic GmbH & Co., KG, Berlin, Germany) for 15 min. PIM-1 was dissolved in chloroform and added to a K$_2$B$_{12}$H$_{12}$ suspension. The resulting solution was stirred with a magnetic bar for a minimum of 15 h, and up to 60 h for a higher loading of IPs. The solution was poured into a leveled circular Teflon® dish, which was covered with glass lead to reduce the chloroform evaporation rate. The slow evaporation of chloroform was ensured by 10 mL/min nitrogen flow through the closed space above the Teflon dish. After solvent evaporation, the prepared membranes were delaminated from the Teflon® surface and conditioned by soaking in methanol for approximately 4 h. Immersing the membranes in methanol reverses prior to film formation history, in a manner similar to protocols previously developed for high free volume polyacetylenes and PIM-1 [28,29]. The methanol-treated membranes were dried in high vacuum for 16 h at 120 °C. The thickness of the membranes was measured by a digital micrometer (Deltascopes MP2C, Helmut Fischer GmbH, Sindelfingen, Germany), ranged between 95 to 101 μm.

4.2. Thermal Gravimetric Analysis (TGA)

Investigation of the thermal stability of the pristine PIM-1, $K_2B_{12}H_{12}$, and PIM1/$K_2B_{12}H_{12}$ MMMs were performed by thermogravimetrical analysis (TGA) on a TG209 F1-Iris instrument from the Netzsch Company (Gerätebau GmbH, Selb, Germany). At least 10 mg of each sample was placed into a sample holder. The experiments were conducted under argon flow (20 mL/min) from 30 to 900 °C with at heating rate 10 K/min.

4.3. Scanning Electron Microscopy (SEM)

A LEO 1550VP instrument (Carl Zeiss AG, Oberkochen, Germany) was used to study the morphology of pure PIM-1 and PIM1/$K_2B_{12}H_{12}$ MMMs, which was equipped with a field emission cathode operated at 1–1.5 kV. Samples for scanning electron microscopy (SEM, Carl Zeiss AG, Oberkochen, Germany) were prepared by freezing the prepared membranes in liquid nitrogen and then breaking them to investigate the homogeneity of the IPs throughout the MMMs and compatibility between the IPs and the polymer phase. The samples were dried overnight in a vacuum oven at 30 °C and then coated with a thin Pt layer using a sputtering device under argon flow.

4.4. Density Measurements

The density of the membranes was determined by the buoyancy method following Equation (8)

$$\rho = \left(\frac{W_A}{W_A - W_L} \right) \rho_L \tag{8}$$

where ρ and ρ_L are the densities of the membranes and perfluorinated liquid (Fluorinert FC 77), respectively, W_A and W_L are the weight of membranes in the air and in perfluorinated liquid, respectively. All the density measurements were done at 26 °C.

4.5. Gas Transport Properties

The permeability of single gases (H_2, O_2, N_2, CH_4, and CO_2) were measured using a constant volume variable pressure time lag apparatus at 30 °C. The permeability (*P*), diffusivity (*D*), solubility (*S*) and selectivity (*α*) for gases *i* and *j* were determined under steady state by the following Equations [30–32]:

$$P = D \times S = \frac{V_p l (p_{p2} - p_{p1})}{ART\Delta t \left[p_f - (p_{p2} + p_{p1}/2) \right]} \tag{9}$$

$$D = \frac{l^2}{6\theta} \tag{10}$$

$$\alpha_{ij} = \frac{P_i}{P_j} = \frac{S_i \times D_i}{S_j \times D_j} \tag{11}$$

where *Vp* is the constant permeate volume, *R* the gas constant, *l* the film thickness, *A* is the effective area of the membrane, Δt is the time for the permeate pressure increase from p_{p1} to p_{p2}, p_f is the feed pressure, and θ is the time-lag. The solution–diffusion transport model [21] was applied to discuss the gas transport properties of PIM-1 and PIM-1 MMMs, and the selectivities of membranes for gas "*i*" relative to another one "*j*", which is the ratio of their permeabilities determined using Equation (6).

4.6. Gas Sorption

Static and dynamic sorption measurements were performed on a magnetic suspension balance (MSB) (Rubotherm GmbH, Bochum, Germany). Static sorption measurements allow the determination of the sorption isotherms, Langmuir hole affinity parameter (b) and the capacity parameter (C'_H) for pristine PIM-1 and PIM1/$K_2B_{12}H_{12}$ MMMs according to Equation (2) [19]. Dynamic sorption measurements can be used to determine the diffusion coefficient of gas in pristine PIM-1 and PIM1/$K_2B_{12}H_{12}$ MMMs by means of Equation (3) [33].

4.6.1. Static Sorption Experiments

The amount of pure gases adsorbed m_{ADS} in the samples (PIM1/$K_2B_{12}H_{12}$ MMMs) was calculated from the volume of the samples (calculated from the density of the samples as determined from the standard buoyancy technique explained in the above section),the gas mass uptake of the samples, and the molar volume and molecular weight of the gas probe. A minimum of 50 mg of sample was used. For each measurement, the samples were evacuated at 353 K for 18 h at $P \leq 10^{-6}$ millibar. All tubing and chambers were also degassed by applying vacuum ($P \leq 10^{-6}$ millibar).The evacuated samples were then cooled down to the specified temperature (303 K) with a ramping rate of 1 K/min. The different used gases have a purity of 99.99% in this measurement. The gravimetric sorption studies in this research were conducted at a temperature of 303 ± 0.1 K and a pressure range of 0.01–8 bar.

4.6.2. Dynamic Sorption Experiments

The diffusion coefficient of gas was calculated using a dynamic sorption experiment for PIM-1 and PIM1/$K_2B_{12}H_{12}$ MMMs. Before the start of each experiment, the thickness of the membrane samples was measured. Prior to pressurization at 1 bar, the sample was evacuated for 18 h. The mass uptake of the sample (M_t) was calculated according to Equation (12):

$$M_t = w_t - \left[w_0 - \left(v_t \times \rho_{gas} \right) \right] \tag{12}$$

where w_0 (g) is the weight of the sample at zero sorption, v_t (cm^3) is the volume of the sample at time t(s) and ρ_{gas} is the density of the gas (g/cm^3). To correct the recorded weight (w_t (g)) for buoyancy effects, the Archimedes principle was used. Subsequently, the ratio of M_t/M_∞ was obtained as a function of time(s). Since in the case of membranes, complete equilibrium could not be established within the time scale of the experiment, in that case, the pseudo-infinite mass uptake after 14 h was used. The obtained data were fitted using Equation (3) to obtain the diffusion coefficients for the membrane samples.

5. Results and Discussion

5.1. Inorganic Particle Characterization

The thermal stability of IPs was investigated by means of TGA. Figure 3 illustrates that no weight loss occurred below 100 °C for both IPs [$K_2B_{12}H_{12}$ and $N(C_4H_9)_4B_{12}H_{12}$], which indicates the absence of residual solvents. $K_2B_{12}H_{12}$ shows no weight loss and remains stable up to the final temperature of 700 °C. For comparison, sample [$N(C_4H_9)_4]B_{12}H_{12}$ shows a large weight loss (~45%) between 200–500 °C. In this temperature range, [$N(C_4H_9)_4]B_{12}H_{12}$ decomposes into gaseous products. From these results, we conclude that the $K_2B_{12}H_{12}$ are thermally stable up to 700 °C. This is relevant for the preparation of MMMs, since heating the polymer matrix above the T_g or T_m can reduce the formation of non-selective voids [34].

Figure 4a shows the SEM image of $K_2B_{12}H_{12}$ with a distinct crystalline structure. The chemical composition of the IPs was analyzed by EDX spectrometer (Carl Zeiss AG, Oberkochen, Germany), which was attached to the SEM image (Figure 4b). The EDX spectra clearly shows the strong signal of potassium (K) and boron (B) in the crystalline structure of $K_2B_{12}H_{12}$.

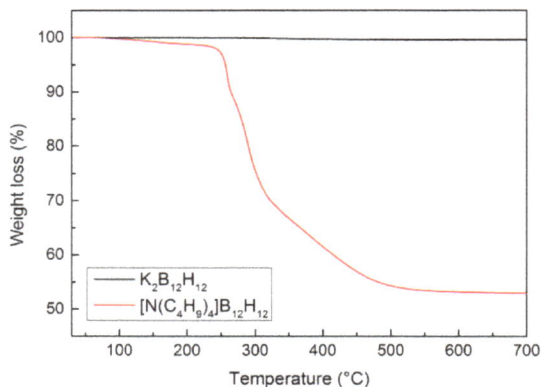

Figure 3. TGA analysis of potassium dodecahydrododecaborate ($K_2B_{12}H_{12}$) and bis-tetrabutyl ammonium *closo*-dodecahydrododecaborate $[N(C_4H_9)_4]_2B_{12}H_{12}$.

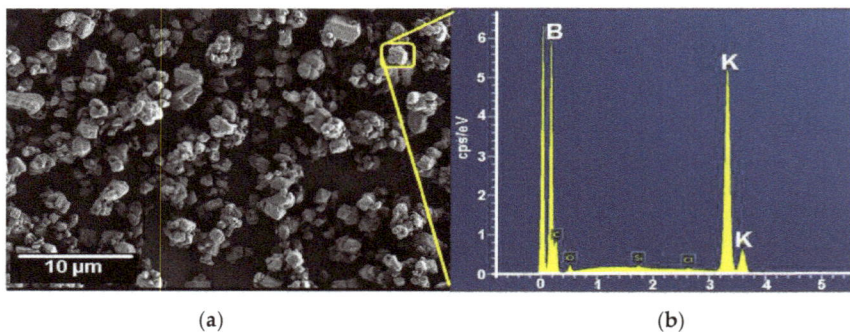

(a)   (b)

Figure 4. SEM image (a) and EDX spectra (b) of $K_2B_{12}H_{12}$.

5.2. Mixed Matrix Membranes (MMMs) Characterization

The effect of temperature on the degradation of pristine PIM-1 and PIM1/$K_2B_{12}H_{12}$ MMMs at a various loading of $K_2B_{12}H_{12}$ is shown in Figure 5. TGA analysis suggests that no residual solvent was present in the films. The PIM1/$K_2B_{12}H_{12}$ MMMs with 5, 10 and 20 wt % loading of $K_2B_{12}H_{12}$ show similar decomposition stages compared to pure PIM-1 and the onset degradation temperature of these samples was observed at 501 ± 2 °C. The higher magnification TGA results of PIM1/$K_2B_{12}H_{12}$ MMMs from 550 to 600 °C were shown the inset Figure 5. Due to the lack of rotational mobility in the backbone of the rigid ladder polymer, it is difficult to observe a glass transition before the degradation of pristine PIM-1 and its MMMs [16].

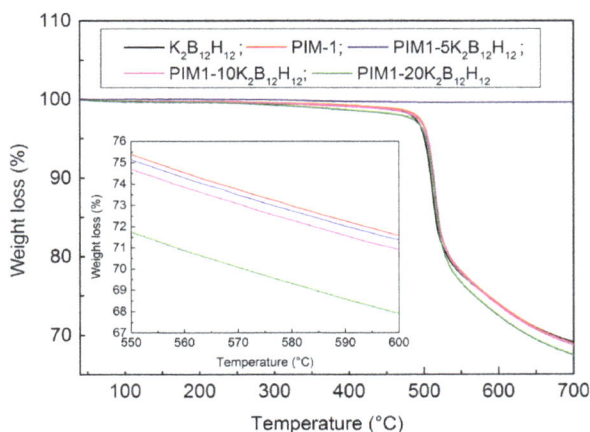

Figure 5. TGA analysis of the pure PIM-1 and PIM1/$K_2B_{12}H_{12}$ MMMs.

Table 1 shows the density and weight loss of various wt % of $K_2B_{12}H_{12}$ in PIM1/$K_2B_{12}H_{12}$ MMMs up to a temperature 650 °C. During TGA analysis, the initial weight loss of the samples was affected by buoyancy, which means that the samples and ceramic pan appeared to gain weight before significant decomposition occurred due to the difference in thermal conductivity, density and heat capacity for the purging gas and the sample [35]. However, the buoyancy effect was less apparent at a higher temperature. Thus, the initial wt % of all the samples was set at 100 °C. The $K_2B_{12}H_{12}$ concentration in the polymer was considered as the volume fraction (ϕ_{IP}), which appears slightly higher than the weight fraction term due to the density difference between $K_2B_{12}H_{12}$ and polymer. PIM-1 and PIM1/$K_2B_{12}H_{12}$ MMMs began weight loss at approximately 501 ± 2 °C. The weight loss up to 700 °C (W_{700}) increased slightly with the addition of $K_2B_{12}H_{12}$.

Table 1. Physical and thermal properties of $K_2B_{12}H_{12}$, PIM-1, and PIM1/$K_2B_{12}H_{12}$ MMMs.

Membrane	Volume Fraction ϕ_{IP} (%)	$K_2B_{12}H_{12}$ Loading (%)	w_{700} (%)	ρ (g/cm^3)
PIM-1	0	0	32.17	1.066
PIM-2.5 $K_2B_{12}H_{12}$	2.58	2.5	32.46	1.078
PIM-5 $K_2B_{12}H_{12}$	5.16	5	33.10	1.077
PIM-10 $K_2B_{12}H_{12}$	10.21	10	33.20	1.072
PIM-20 $K_2B_{12}H_{12}$	20.53	20	34.43	1.067
$K_2B_{12}H_{12}$	-	-	0.6	1.031 *

* determined from Micromeritics AccuPyc 1330 pycnometer. w_{700}: weight loss up to 700 °C. ρ: density of membrane.

The optical transparencies of PIM-1 and PIM1/$K_2B_{12}H_{12}$ MMMs are shown in Figure 6. These images confirm the improved dispersion of the inorganic particles up to 10 wt % loading. At higher filler content (20 wt %), there is greater agglomeration of inorganic particles in the polymer matrix (see PIM-20 $K_2B_{12}H_{12}$ MMMs film in Figure 6). PIM-1 and PIM1/$K_2B_{12}H_{12}$ MMMs films were more flexible and mechanically stable. The mechanical stability deteriorated beyond 20 wt % filler content in the polymer.

Figure 7 shows the cross-sectional SEM images of PIM-1 and PIM1/$K_2B_{12}H_{12}$ MMMs at different $K_2B_{12}H_{12}$ loadings. $K_2B_{12}H_{12}$ tend to be well-distributed throughout the membrane surface with a 5 and 10 wt % $K_2B_{12}H_{12}$ loading. (Figure 7b,c). As the $K_2B_{12}H_{12}$ loadings were further increased to 20 wt %, the $K_2B_{12}H_{12}$ started to agglomerate throughout the PIM-1 matrix (see Figure 6). Figure 7a–d shows highly-magnified images of the PIM-1 and PIM1/$K_2B_{12}H_{12}$ MMMs incorporated with 5, 10 and 20 wt % of $K_2B_{12}H_{12}$ (showed by a yellow circle). From this observation and optical images (see Figure 6), we can conclude that the threshold limit for the addition of $K_2B_{12}H_{12}$ into the polymer matrix to prevent agglomeration is typically around 20 wt % and the optimum for the addition of $K_2B_{12}H_{12}$ is lower than 20 wt %.

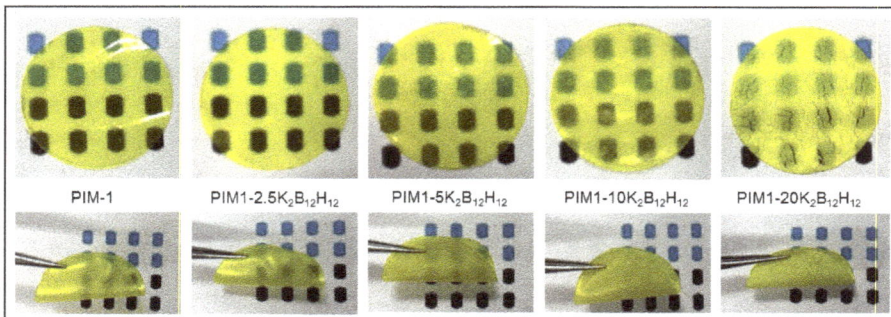

Figure 6. Optical images of PIM-1 and PIM1/$K_2B_{12}H_{12}$ MMMs.

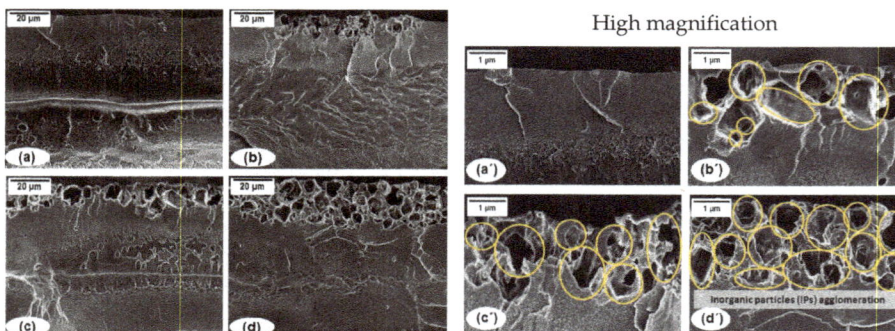

Figure 7. Cross section SEM images of (**a,a′**) PIM-1, PIM1/$K_2B_{12}H_{12}$ MMMs incorporated with (**b,b′**) 5 wt %, (**c,c′**) 10 wt % and (**d,d′**) 20 wt % of $K_2B_{12}H_{12}$.

5.3. Gas Permeation Properties

5.3.1. Effects of $K_2B_{12}H_{12}$ Content on PIM1/$K_2B_{12}H_{12}$ MMM Gas Separation Performance

In order to systematically study the effect of $K_2B_{12}H_{12}$ loading on the PIM1/$K_2B_{12}H_{12}$ MMM gas separation performance, MMMs were fabricated with different wt % incorporation of $K_2B_{12}H_{12}$. The permeability results of PIM1/$K_2B_{12}H_{12}$ MMMs for H_2, O_2, N_2, CO_2 and CH_4 gases are shown in Table 2. The order of gas permeability was observed as $CO_2 > H_2 > O_2 > CH_4 > N_2$. The addition of 2.5 wt % of $K_2B_{12}H_{12}$ loading to the polymer matrix resulted in a 3% increase in the permeability of H_2, while the permeability of N_2, O_2, CO_2 and CH_4 increased 16%, 10%, 17%, and 23%, respectively. Furthermore, a significant enhancement in permeability as a function of $K_2B_{12}H_{12}$ loading in the polymer matrix was observed between 5 to 10 wt %. From the previous report on the permeation enhancement of MMMs [5], these results suggest that the interaction between polymer-chain segments and IPs may disrupt the polymer-chain packing and thus enhance the gas diffusion due to more free volume introduced among the polymer chains and defects at the polymer/IP interface. The permeability of gas molecules such as H_2, N_2, O_2, CO_2 and CH_4 decreases as $K_2B_{12}H_{12}$ loading increased from 10 to 20 wt % in the polymer matrix. Some agglomerates form in the polymer matrix at high loading

(20 wt %), which may decrease the total free volume and tortuosity around the agglomerated $K_2B_{12}H_{12}$ domains, leading to a slight deterioration of the permeation.

Table 2. Gas permeabilities of various gases in pure PIM-1 and PIM1/$K_2B_{12}H_{12}$ MMMs.

Membrane	Permeability (Barrer)				
	H_2	N_2	O_2	CO_2	CH_4
PIM-1	3274 ± 5	483 ± 10	1396 ± 13	9896 ± 28	789 ± 15
PIM1-2.5 $K_2B_{12}H_{12}$ MMM	3347 ± 8 (3%)	562 ± 11 (16%)	1539 ± 12 (10%)	11598 ± 20 (17%)	974 ± 18 (23%)
PIM1-5 $K_2B_{12}H_{12}$ MMM	3707 ± 9 (13%)	641 ± 10 (33%)	1675 ± 11 (20%)	12036 ± 21 (22%)	1148 ± 16 (45%)
PIM1-10 $K_2B_{12}H_{12}$ MMM	4025 ± 8 (22%)	772 ± 9 (60%)	1831 ± 14 (31%)	12954 ± 23 (31%)	1436 ± 16 (82%)
PIM1-20 $K_2B_{12}H_{12}$ MMM	3436 ± 7 (5%)	607 ± 12 (25%)	1600 ± 14 (14%)	11729 ± 23 (18%)	1123 ± 14 (42%)

(% increment from pure polymer).

In addition, the gas permeabilities of PIM-1 containing $K_2B_{12}H_{12}$ were higher than pure PIM-1 and increasing up to the optimum limit. This trend is clearly depicted in Figure 8, which presents the normalized permeability of PIM1/$K_2B_{12}H_{12}$ MMMs for O_2, N_2, CH_4, and CO_2 gases as a function of $K_2B_{12}H_{12}$ volume fraction (ϕ_{IP}).

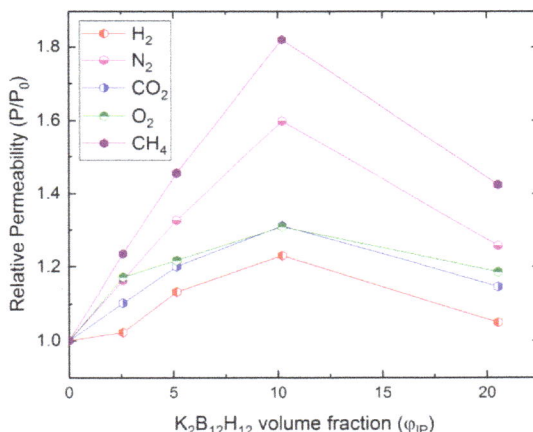

Figure 8. Relative permeability (i.e., ratio of permeability of PIM1/$K_2B_{12}H_{12}$ with pure polymer PIM-1) of PIM1/$K_2B_{12}H_{12}$ MMMs to a variety of gas penetrates as a function of $K_2B_{12}H_{12}$ volume fraction (ϕ_{IP}).

Table 3 shows the ideal separation factors for pure PIM-1 and PIM1/$K_2B_{12}H_{12}$ MMMs. At 2.5–10 wt % $K_2B_{12}H_{12}$ loading, the permselectivity was found to be decreased compared to the pure PIM-1. However, the selectivity increased at 20 wt % $K_2B_{12}H_{12}$ loading due to a significant decrease in permeability. It is shown in Table 3 that the CH_4/N_2 separation factor increased as the amount of $K_2B_{12}H_{12}$ increased due to the higher adsorption capacity for CH_4 over N_2. Despite increases in the permeability of O_2 and N_2, the O_2/N_2 separation factor remained virtually unchanged because $K_2B_{12}H_{12}$ were not selective for either O_2 or N_2. In addition, the constant O_2/N_2 separation factor in PIM1/$K_2B_{12}H_{12}$ MMMs suggests that the prepared membranes do not have any unselective voids at the polymer/$K_2B_{12}H_{12}$ interface.

Recently, various trends of MMMs in terms of relative trade-off in permeability and permselectivity have been noted. Many permselectivity increments were seen with the addition of activated carbon, fused silica and metal organic frameworks (MOF) [36]. The gas separation performance of PIM1/$K_2B_{12}H_{12}$ MMMs was plotted on a Robeson upper bound plot in order to compare the results with the literature data. The Figure 9 shows the Robeson upper bound 2008 [37] for CO_2/N_2 gas pairs and the results of these MMMs with different filler content. The incorporation of fillers in the PIM-1 polymer increases the efficiency of this membrane type in the separation of CO_2 gas over N_2.

Table 3. Selectivity of various gas pairs for pure PIM-1 and PIM1/$K_2B_{12}H_{12}$ MMMs.

Membrane	Permselectivity					
	H_2/N_2	H_2/CH_4	CH_4/N_2	O_2/N_2	CO_2/N_2	CO_2/CH_4
PIM-1	6.8	4.2	1.6	2.9	20.5	12.5
PIM1-2.5 $K_2B_{12}H_{12}$ MMM	6.0	3.4	1.7	2.7	20.7	11.9
PIM1-5 $K_2B_{12}H_{12}$ MMM	5.8	3.2	1.8	2.6	18.8	10.5
PIM1-10 $K_2B_{12}H_{12}$ MMM	5.2	2.8	1.9	2.4	16.8	9.0
PIM1-20 $K_2B_{12}H_{12}$ MMM	5.6	3.0	1.8	2.6	19.3	10.4

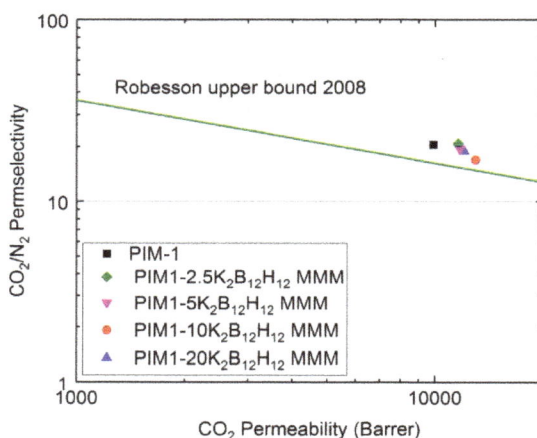

Figure 9. Trade-off between CO_2 permeability and CO_2/N_2 permselectivity of PIM-1 and PIM1/$K_2B_{12}H_{12}$ MMMs relative to Robeson upper bound plot.

5.3.2. Influence of Temperature on the Gas Separation Performance of PIM1/$K_2B_{12}H_{12}$ MMMs

Temperature effects on PIM1/$K_2B_{12}H_{12}$ MMMs were studied over a temperature range of 283–343 K (10, 30, 50 and 70 °C) for single gas at one bar feed pressure. Figure 10 shows the permeability of N_2, CH_4, CO_2, and O_2 for PIM-1 and PIM1/$K_2B_{12}H_{12}$ MMMs as a function of the inverse absolute temperature. From Figure 10, it can be seen that the permeability of N_2 and CH_4 increased with increasing temperature, while for CO_2 and O_2, the permeability decreased with increasing temperature. This result indicates that highly sorbed gases like CO_2 do not affect the permeation rate of lighter gases in subsequent runs [15,38]. However, a careful examination shows that the permeability of all gases is higher in 2.5–10 wt % than 20 wt % PIM1/$K_2B_{12}H_{12}$ MMMs and the pristine PIM-1 membrane at each temperature.

Figure 11 shows the O_2/N_2, CO_2/N_2 and CO_2/CH_4 selectivity of the pure PIM-1 and PIM1/$K_2B_{12}H_{12}$ MMMs as a function of the inverse of absolute temperature. It was observed that the selectivity for a given gas pair decreases with an increase in the temperature of pure PIM-1 and PIM1/$K_2B_{12}H_{12}$ MMMs. Hence, the incorporation of $K_2B_{12}H_{12}$ does not change any selectivity pattern at a higher temperature. However, a significant difference in selectivity at a lower temperature was observed for PIM-1. It shows that the O_2/N_2 selectivity at 333 K is nearly 2.7; at 283 K it reaches 4.6, while for CO_2/N_2 selectivity is around 30.9 at low temperature and 14.1 at elevated temperature 343 K (70 °C).

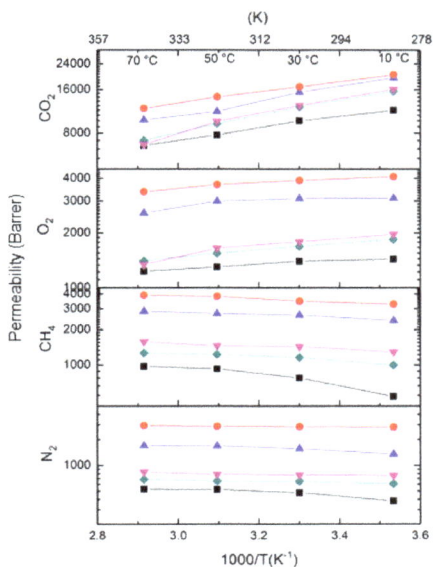

Figure 10. Permeability of N_2, CH_4, CO_2 and O_2 in PIM-1 and PIM1/$K_2B_{12}H_{12}$ MMMs as a function of reciprocal temperature ((■-black) PIM-1, (♦-olive) 2.5 wt % PIM1/$K_2B_{12}H_{12}$ MMM, (▼-pink) 5 wt % PIM1/$K_2B_{12}H_{12}$ MMM, (●-red) 10 wt % PIM1/$K_2B_{12}H_{12}$ MMM, (▲-blue) 20 wt % PIM1/$K_2B_{12}H_{12}$ MMM).

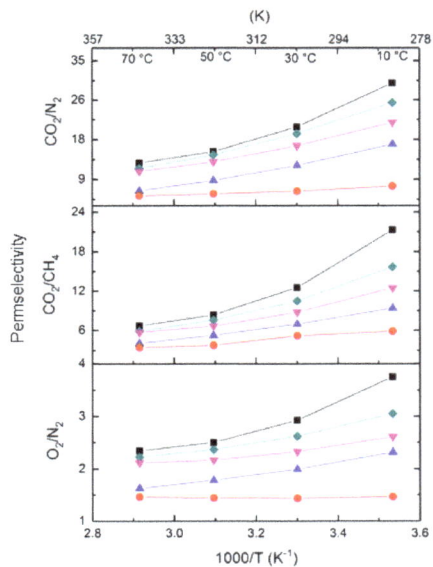

Figure 11. Selectivity of O_2/N_2, CO_2/N_2 and CO_2/CH_4 in PIM-1 and PIM1/$K_2B_{12}H_{12}$ MMMs as a function of reciprocal temperature ((■-black) PIM-1, (♦-olive) 2.5 wt % PIM1/$K_2B_{12}H_{12}$ MMM, (▼-pink) 5 wt % PIM1/$K_2B_{12}H_{12}$ MMM, (●-red) 10 wt % PIM1/$K_2B_{12}H_{12}$ MMM, (▲-blue) 20 wt % PIM1/$K_2B_{12}H_{12}$ MMM).

In order to understand the temperature dependence of N_2, O_2, CO_2 and CH_4 permeabilities in PIM1/$K_2B_{12}H_{12}$ MMMs, the data were correlated with the Arrhenius equation and the activation energy of permeation (E_P) was determined using the following relationship:

$$P = P_0 \, exp\left(\frac{-E_P}{RT}\right) \qquad (13)$$

where P is the gas permeability, P_0 is the pre-exponential factor, (E_P) is the activation energy of permeation (J/mol), R is the gas constant (8.314 J/(mol·K)) and T is the absolute temperature. The given equation was valid in a temperature range that does not cause significant thermal transitions in the polymer. Table 4 shows the activation energy of permeation (E_P) of PIM-1 and PIM1/$K_2B_{12}H_{12}$ MMMs, which were determined from the slope of the Arrhenius plot.

According to the literature, the activation energy of permeation was the sum of the activation energy of diffusion (E_D), and the enthalpy of sorption (ΔH_S),

$$E_P = E_D + \Delta H_S \qquad (14)$$

Table 4. Activation energy of permeation for pristine PIM-1 and PIM1/$K_2B_{12}H_{12}$ MMMs.

Membrane	E_P (kJ/mol)	
	N_2	CO_2
PIM-1	18.5	−3.3
PIM1-2.5 $K_2B_{12}H_{12}$ MMM	13.7	−4.0
PIM1-5 $K_2B_{12}H_{12}$ MMM	6.4	−4.6
PIM1-10 $K_2B_{12}H_{12}$ MMM	5.5	−5.0
PIM1-20 $K_2B_{12}H_{12}$ MMM	2.4	−3.1

Generally, the gas permeability of all conventional glassy polymers increases with increased temperature, because $E_D + H_S > 0$ and $|E_D| / |H_S| > 1$. An exception to this rule is the temperature dependence of gas permeability in high free volume polymers such as PIM-1, i.e., gas permeabilities decrease with increase temperature for condensable gas (e.g., CO_2), where $|E_D| / |H_S| < 1$ [15]. Therefore, the negative activation energies of permeation in PIM-1 and PIM1/$K_2B_{12}H_{12}$ MMMs result from very small activation energies of diffusion, which indicates that the dependence of permeability on temperature is much weaker. In addition, the negative value of E_P is characteristic of the decrease of CO_2 permeability with the increase of temperature, which was clearly observed in Figure 10. Another case, the N_2 permeability of PIM-1 and PIM1/$K_2B_{12}H_{12}$ MMMs, was strongly temperature-dependent and E_P values were the same order of magnitude as those of conventional glassy polymers. Moreover, negative E_P was observed for microporous solids in which the pore dimensions were relatively larger than the diffusing gas molecules [39].

5.4. Gas Sorption

5.4.1. Static Gas Sorption

Static gas sorption measurements were performed to characterize the sorption behavior of pure PIM-1 and PIM1/$K_2B_{12}H_{12}$ MMMs. Figure 12 represents N_2, O_2, and CH_4 sorption isotherms in PIM-1 and PIM-1 containing 2.5, 5, 10 and 20 wt % $K_2B_{12}H_{12}$ at 303 K. From Figure 12, the sorption of N_2 and O_2 was much less than that of other gases, such as CO_2 and CH_4, owing to their lower condensability and weak interaction with the PIM-1 polymer. On the other hand, the sorption curve concave to the pressure axis was observed for CO_2 and CH_4, this was a general trend for glassy polymers and can be described by the so-called dual-mode sorption model [40,41]. The amount of gas absorbed in PIM1/$K_2B_{12}H_{12}$ MMMs films for each gas depends on the $K_2B_{12}H_{12}$ content, as shown in Figure 12.

The relative increase in gas absorption was small at 20 wt % of $K_2B_{12}H_{12}$ loading (showing more discrepancy in sorption measurement—see Figure 12) in comparison to the increases seen at 2.5, 5 and 10 wt % of $K_2B_{12}H_{12}$ loading. Therefore, the presence of $K_2B_{12}H_{12}$ increases the relative sorption of gases in the membrane, at higher $K_2B_{12}H_{12}$ contents; this increase could be constrained polymer chain packing at the $K_2B_{12}H_{12}$/polymer interface.

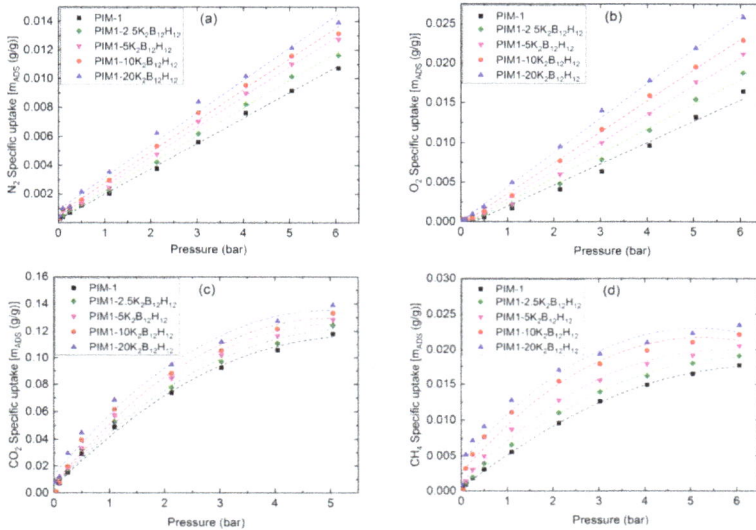

Figure 12. (**a**) N_2; (**b**) O_2; (**c**) CO_2; and (**d**) CH_4 adsorption isotherm in PIM-1 and PIM1/$K_2B_{12}H_{12}$ MMMs (dashed lines represent the fitting curve).

When both sorption isotherms of CO_2 and CH_4 were fitted with the dual-mode sorption model (Equation (2)), Henry's constants (k_D), the Langmuir capacity constants (C'_H) and the Langmuir affinity constants (b) can be obtained using a non-linear regression method and these were shown in Table 5. The low Henry constants for both CO_2 and CH_4 indicate that the major sorption mechanism inside PIM-1 was Langmuir sorption, which takes place in the non-equilibrium excess volume occurring in glassy polymers [42]. The addition of IPs to the polymer matrix could affect and possibly disturb or alter this excess volume. When the dual mode sorption parameters for different wt % of $K_2B_{12}H_{12}$ are compared in Table 5, an increase with increasing $K_2B_{12}H_{12}$ loading was visible for all parameters for both CO_2 and CH_4. This implies that the addition of $K_2B_{12}H_{12}$ increases the maximum sorption capacity and the affinity towards CO_2 and CH_4, but does not provide any additional sorption selectivity, since the ideal sorption selectivity does not increase.

The maximum sorption capacity, C'_H, in 20 wt % PIM1/$K_2B_{12}H_{12}$ MMMs was decreased by 5% and 2% for CO_2 and CH_4, respectively, compared to 10 wt % PIM1/$K_2B_{12}H_{12}$ MMMs. This difference can be explained by sorption limitations in the $K_2B_{12}H_{12}$ particles due to the surrounding polymer. From the SEM images in Figure 7c,c', there was large area of agglomeration between the polymer matrix and the $K_2B_{12}H_{12}$ at 20 wt % loading, which might reduce the sorption capacity on the outside of the $K_2B_{12}H_{12}$, where interaction with the polymer takes place, or limits the diffusion into $K_2B_{12}H_{12}$. Moreover, the addition of $K_2B_{12}H_{12}$ particles might have an influence on the diffusion coefficient, which is discussed in the next paragraph.

Table 5. Fitted dual-mode sorption parameters in PIM-1 and PIM1/$K_2B_{12}H_{12}$ MMMs of CO_2 and CH_4 sorption isotherm.

Feed Gas	$K_2B_{12}H_{12}$ Loading (wt %)	Dual Mode Sorption Model Parameter		
		k_D	C'_H	b
	0	2.330	104.630	0.415
	2.5	2.440	105.830	0.422
CO_2	5	2.530	113.750	0.440
	10	2.600	120.105	0.491
	20	2.58	115.81	0.444
	0	0.581	62.097	0.135
	2.5	0.59	63.957	0.141
CH_4	5	0.604	65.042	0.154
	10	0.627	66.741	0.167
	20	0.611	65.412	0.151

Units of k_D = cm^3(STP)/(cm^3·atm)$_{polymer}$, C'_H = cm^3(STP)/cm^3 $_{polymer}$, b = atm^{-1}.

5.4.2. Dynamic Gas Sorption

Dynamic sorption experiments were performed to determine the kinetic diffusion coefficients of the PIM-1 and PIM1/$K_2B_{12}H_{12}$ MMMs with various wt % of $K_2B_{12}H_{12}$. It was important to verify whether all fitting parameters can be accurately obtained with the given film thickness. Figure 13 depicts the CO_2 kinetic sorption fractional uptake curves in PIM-1 and PIM-1 containing 5 wt %, 10 wt % and 20 wt % $K_2B_{12}H_{12}$. From Figure 13, CO_2 uptake kinetics was normalized to account for differences in film thickness, and the sorption equilibrium was attained much more rapidly in PIM1/$K_2B_{12}H_{12}$ MMMs than pure PIM-1. This result implies faster diffusion in PIM1/$K_2B_{12}H_{12}$ MMMs, which is qualitatively consistent with the concentration-averaged diffusion coefficients.

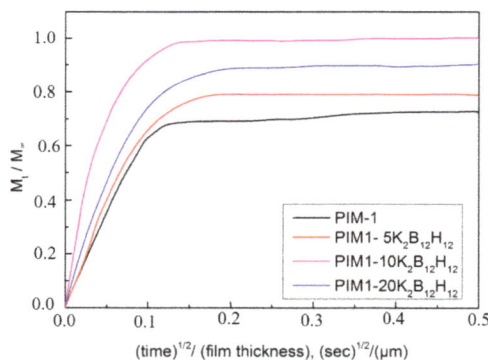

Figure 13. CO_2 kinetic uptake curves in PIM-1 and PIM1/$K_2B_{12}H_{12}$ MMMs at 303 K and 1 bar.

In addition to the diffusion coefficient (D) that was calculated from steady-state transport data, diffusion coefficients may also be estimated from the dynamic sorption. Kinetic or transient diffusion coefficients, D, were extracted from the data in Figure 13 by application of the one-dimensional form of following Equation (15), which was modified from Equation (3) for Fick's diffusion law:

$$\frac{M_t}{M_\infty} = 4\left(\frac{Dt}{\pi l^2}\right)^{1/2} \tag{15}$$

where M_t was the mass gain (by the polymer film) at time t, M_∞ is the maximum mass gain, D was the diffusivity gas penetrant and l was the thickness of the film.

Figure 14 shows CO_2 diffusion coefficients in PIM-1 and $PIM1/K_2B_{12}H_{12}$ MMMs determined from the kinetic sorption studies (using Equation (15)). The diffusion coefficients were calculated from the time-lag method, included as well in Figure 14 for comparison.

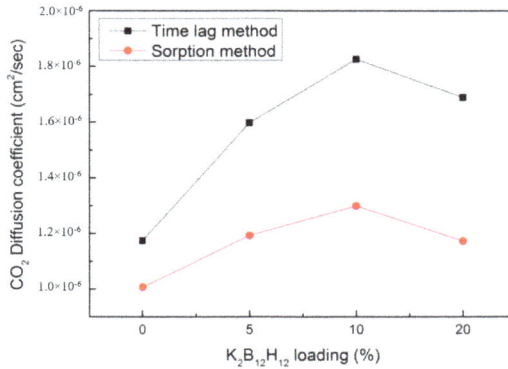

Figure 14. CO_2 diffusion coefficient in PIM-1 and $PIM1/K_2B_{12}H_{12}$ MMMs.

Although the absolute values of the diffusion coefficients obtained by the time-lag and kinetic sorption methods were different [43], qualitatively the changes were consistent. Typically, the kinetic diffusion coefficients measured by gravimetric sorption were lower than those obtained by the time-lag. This discrepancy in results was observed because kinetic (transient) uptake experiments involve additional diffusion into the dead-end pores, while transport through dead-end pores does not play a role in steady-state permeation (time-lag) experiments [44].

6. Conclusions

Mixed matrix membranes were prepared successfully adding different amounts of $K_2B_{12}H_{12}$ as IPs into a PIM-1 as a polymer matrix. The prepared $PIM1/K_2B_{12}H_{12}$ MMMs were characterized by scanning electron microscopy (SEM), thermogravimetric analysis (TGA), single gas permeation tests and sorption measurement. $K_2B_{12}H_{12}$ were well-dispersed in the polymer matrix at a loading of 2.5, 5, and 10 wt %, while at 20 wt % the $K_2B_{12}H_{12}$ forms agglomeration and phase separation in the polymer matrix, which was confirmed by SEM and optical images. The permeability performance of the prepared $PIM1/K_2B_{12}H_{12}$ MMMs mainly depends on the addition of IPs rather than the effect of the interfacial zone because the O_2/N_2 gas pair selectivity was constant for all MMMs. Overall increases in gas permeability and diffusivity were observed for all tested gases, suggesting that IPs could disrupt more polymer chain packing. The sorption isotherms in PIM-1 and $PIM1/K_2B_{12}H_{12}$ MMMs exhibited typical dual-mode sorption behaviors for the gases CO_2 and CH_4. The CO_2 diffusion coefficient calculated by the dynamic sorption method was lower than the time-lag method for PIM-1 and $PIM1/K_2B_{12}H_{12}$ MMMs. This is the first report of the gas transport performance of MMMs using $K_2B_{12}H_{12}$ and a PIM-1 polymer. It is clear that the addition of $K_2B_{12}H_{12}$ to a polymer matrix can improve certain gas pair selectivities, as well as the permeability of small gas molecules.

Acknowledgments: This work was financially supported by the Helmholtz Association of German Research Centres through the project Helmholtz Portfolio MEM-BRAIN. The authors thank Silvio Neumann for the polymer synthesis, Sofi Dami and Clarissa Abetz for SEM measurements and Jelena Lillepärg for sorption measurement.

Author Contributions: Muntazim Munir Khan carried out the experiments and drafted the manuscript. Volkan Filiz supervised the study and Sergey Shishatskiy was involved in scientific discussions. All authors read and approved the final manuscript.

Conflicts of Interest: The authors declare that they have no conflicts of interest.

References

1. Freeman, B.D. Basis of permeability/selectivity tradeoff relations in polymeric gas separation membranes. *Macromolecules* **1999**, *32*, 375–380. [CrossRef]
2. Robeson, L.M. Correlation of separation factor versus permeability for polymeric membranes. *J. Membr. Sci.* **1991**, *62*, 165–185. [CrossRef]
3. Cornelius, C.J.; Marand, E. Hybrid silica-polyimide composite membranes: Gas transport properties. *J. Membr. Sci.* **2002**, *202*, 97–118. [CrossRef]
4. Mahajan, R.; Burns, R.; Schaeffer, M.; Koros, W.J. Challenges in forming successful mixed matrix membranes with rigid polymeric materials. *J. Appl. Polym. Sci.* **2002**, *86*, 881–890. [CrossRef]
5. Bushell, A.F.; Attfield, M.P.; Mason, C.R.; Budd, P.M.; Yampolskii, Y.; Starannikova, L.; Rebrov, A.; Bazzarelli, F.; Bernardo, P.; Carolus Jansen, J.; et al. Gas permeation parameters of mixed matrix membranes based on the polymer of intrinsic microporosity pim-1 and the zeolitic imidazolate framework zif-8. *J. Membr. Sci.* **2013**, *427*, 48–62. [CrossRef]
6. Ismail, A.F.; Rahim, R.A.; Rahman, W.A.W.A. Characterization of polyethersulfone/matrimid® 5218 miscible blend mixed matrix membranes for o2/n2 gas separation. *Sep. Purif. Technol.* **2008**, *63*, 200–206. [CrossRef]
7. Safronov, A.V.; Jalisatgi, S.S.; Lee, H.B.; Hawthorne, M.F. Chemical hydrogen storage using polynuclear borane anion salts. *Int. J. Hydrogen Energy* **2011**, *36*, 234–239. [CrossRef]
8. Stibr, B. Carboranes other than C2B10H12. *Chem. Rev.* **1992**, *92*, 225–250. [CrossRef]
9. Bregadze, V.I.; Sivaev, I.B.; Glazun, S.A. Polyhedral boron compounds as potential diagnostic and therapeutic antitumor agents. *Anti-Cancer Agents Med. Chem. Anti-Cancer Agents* **2006**, *6*, 75–109. [CrossRef]
10. Li, T.; Jalisatgi, S.S.; Bayer, M.J.; Maderna, A.; Khan, S.I.; Hawthorne, M.F. Organic syntheses on an icosahedral borane surface: Closomer structures with twelvefold functionality. *J. Am. Chem. Soc.* **2005**, *127*, 17832–17841. [CrossRef] [PubMed]
11. Plesek, J. Potential applications of the boron cluster compounds. *Chem. Rev.* **1992**, *92*, 269–278. [CrossRef]
12. Sivaev, I.B.; Bregadze, V.V. Polyhedral boranes for medical applications: Current status and perspectives. *Eur. J. Inorg. Chem.* **2009**, *2009*, 1433–1450. [CrossRef]
13. Drissner, D.; Kunze, G.; Callewaert, N.; Gehrig, P.; Tamasloukht, M.B.; Boller, T.; Felix, G.; Amrhein, N.; Bucher, M. Lyso-phosphatidylcholine is a signal in the arbuscular mycorrhizal symbiosis. *Science* **2007**, *318*, 265–268. [CrossRef] [PubMed]
14. Budd, P.M.; Ghanem, B.S.; Makhseed, S.; McKeown, N.B.; Msayib, K.J.; Tattershall, C.E. Polymers of intrinsic microporosity (PIMs): Robust, solution-processable, organic nanoporous materials. *Chem. Commu.* **2004**, 230–231. [CrossRef] [PubMed]
15. Budd, P.M.; McKeown, N.B.; Ghanem, B.S.; Msayib, K.J.; Fritsch, D.; Starannikova, L.; Belov, N.; Sanfirova, O.; Yampolskii, Y.; Shantarovich, V. Gas permeation parameters and other physicochemical properties of a polymer of intrinsic microporosity: Polybenzodioxane PIM-1. *J. Membr. Sci.* **2008**, *325*, 851–860. [CrossRef]
16. Budd, P.M.; Msayib, K.J.; Tattershall, C.E.; Ghanem, B.S.; Reynolds, K.J.; McKeown, N.B.; Fritsch, D. Gas separation membranes from polymers of intrinsic microporosity. *J. Membr. Sci.* **2005**, *251*, 263–269. [CrossRef]
17. McKeown, N.B.; Budd, P.M. Polymers of intrinsic microporosity (PIMS): Organic materials for membrane separations, heterogeneous catalysis and hydrogen storage. *Chem. Soc. Rev.* **2006**, *35*, 675–683. [CrossRef] [PubMed]
18. Koros, W.J.; Chern, R.T. *Handbook of Separation Process Technology*; John Wiley & Sons: Hoboken, NJ, USA, 1987.
19. Story, B.J.; Koros, W.J. Comparison of three models for permeation of CO_2/CH_4 mixtures in poly(phenylene oxide). *J. Polym. Sci. B Polym. Phys.* **1989**, *27*, 1927–1948. [CrossRef]
20. Crank, J. *The Mathematics of Diffusion*; Oxford Press: London, UK, 1990.
21. Wijmans, J.G.; Baker, R.W. The solution-diffusion model: A review. *J. Membr. Sci.* **1995**, *107*, 1–21. [CrossRef]
22. Yampolskii, Y.; Pinnau, I.; Freeman, B.D. *Material Science of Membranes*; John Wiley & Sons: Chichester, UK, 2007.
23. Fritsch, D.; Bengtson, G.; Carta, M.; McKeown, N.B. Synthesis and gas permeation properties of spirobischromane-based polymers of intrinsic microporosity. *Macromol. Chem. Phys.* **2011**, *212*, 1137–1146. [CrossRef]
24. Khan, M.M.; Bengtson, G.; Shishatskiy, S.; Gacal, B.N.; Mushfequr Rahman, M.; Neumann, S.; Filiz, V.; Abetz, V. Cross-linking of polymer of intrinsic microporosity (pim-1) via nitrene reaction and its effect on gas transport property. *Eur. Polym. J.* **2013**, *49*, 4157–4166. [CrossRef]

25. Khan, M.; Filiz, V.; Bengtson, G.; Shishatskiy, S.; Rahman, M.; Abetz, V. Functionalized carbon nanotubes mixed matrix membranes of polymers of intrinsic microporosity for gas separation. *Nanoscale Res. Lett.* **2012**, *7*, 504. [CrossRef] [PubMed]

26. Khan, M.; Filiz, V.; Emmler, T.; Abetz, V.; Koschine, T.; Rätzke, K.; Faupel, F.; Egger, W.; Ravelli, L. Free volume and gas permeation in anthracene maleimide-based polymers of intrinsic microporosity. *Membranes* **2015**, *5*, 214–227. [CrossRef] [PubMed]

27. Khan, M.M.; Bengtson, G.; Neumann, S.; Rahman, M.M.; Abetz, V.; Filiz, V. Synthesis, characterization and gas permeation properties of anthracene maleimide-based polymers of intrinsic microporosity. *RSC Adv.* **2014**, *4*, 32148–32160. [CrossRef]

28. Hill, A.J.; Pas, S.J.; Bastow, T.J.; Burgar, M.I.; Nagai, K.; Toy, L.G.; Freeman, B.D. Influence of methanol conditioning and physical aging on carbon spin-lattice relaxation times of poly(1-trimethylsilyl-1-propyne). *J. Membr. Sci.* **2004**, *243*, 37–44. [CrossRef]

29. Nagai, K.; Toy, L.G.; Freeman, B.D.; Teraguchi, M.; Masuda, T.; Pinnau, I. Gas permeability and hydrocarbon solubility of poly[1-phenyl-2-[p-(triisopropylsilyl)phenyl]acetylene]. *J. Polym. Sci. B Polym. Phys.* **2000**, *38*, 1474–1484. [CrossRef]

30. Rahman, M.M.; Filiz, V.; Shishatskiy, S.; Abetz, C.; Neumann, S.; Bolmer, S.; Khan, M.M.; Abetz, V. Pebax® with peg functionalized poss as nanocomposite membranes for CO$_2$ separation. *J. Membr. Sci.* **2013**, *437*, 286–297. [CrossRef]

31. Rahman, M.M.; Filiz, V.; Shishatskiy, S.; Neumann, S.; Khan, M.M.; Abetz, V. Peg functionalized poss incorporated pebax nanocomposite membranes. *Procedia Eng.* **2012**, *44*, 1523–1526. [CrossRef]

32. Shishatskii, A.M.; Yampol'skii, Y.P.; Peinemann, K.V. Effects of film thickness on density and gas permeation parameters of glassy polymers. *J. Membr. Sci.* **1996**, *112*, 275–285. [CrossRef]

33. Macdonald, D.D. The mathematics of diffusion. In *Transient Techniques in Electrochemistry*; Springer: Boston, MA, USA, 1977; pp. 47–67.

34. Moore, T.T.; Koros, W.J. Non-ideal effects in organic–inorganic materials for gas separation membranes. *J. Mol. Struct.* **2005**, *739*, 87–98. [CrossRef]

35. Byoyancy Phenomenon in TGA System, Thermal Analysis and Surface Solution Gmbh (Thermo Electron Corporation). Available online: http://www.thass.org/ (accessed on 29 December 2017).

36. Rezakazemi, M.; Ebadi Amooghin, A.; Montazer-Rahmati, M.M.; Ismail, A.F.; Matsuura, T. State-of-the-art membrane based CO$_2$ separation using mixed matrix membranes (MMMs): An overview on current status and future directions. *Prog. Polym. Sci.* **2014**, *39*, 817–861. [CrossRef]

37. Robeson, L.M. The upper bound revisited. *J. Membr. Sci.* **2008**, *320*, 390–400. [CrossRef]

38. Khan, M.M.; Filiz, V.; Bengtson, G.; Shishatskiy, S.; Rahman, M.M.; Lillepaerg, J.; Abetz, V. Enhanced gas permeability by fabricating mixed matrix membranes of functionalized multiwalled carbon nanotubes and polymers of intrinsic microporosity (PIM). *J. Membr. Sci.* **2013**, *436*, 109–120. [CrossRef]

39. Pinnau, I.; Toy, L.G. Gas and vapor transport properties of amorphous perfluorinated copolymer membranes based on 2,2-bistrifluoromethyl-4,5-difluoro-1,3-dioxole/tetrafluoroethylene. *J. Membr. Sci.* **1996**, *109*, 125–133. [CrossRef]

40. Koros, W.J.; Chan, A.H.; Paul, D.R. Sorption and transport of various gases in polycarbonate. *J. Membr. Sci.* **1977**, *2*, 165–190. [CrossRef]

41. Wang, R.; Cao, C.; Chung, T.-S. A critical review on diffusivity and the characterization of diffusivity of 6FDA–6FPDA polyimide membranes for gas separation. *J. Membr. Sci.* **2002**, *198*, 259–271. [CrossRef]

42. Paul, D.R. Gas sorption and transport in glassy polymers. *Ber. Bunsenges. Phys. Chem.* **1979**, *83*, 294–302. [CrossRef]

43. Merkel, T.C.; He, Z.; Pinnau, I.; Freeman, B.D.; Meakin, P.; Hill, A.J. Sorption and transport in poly(2,2-bis(trifluoromethyl)-4,5-difluoro-1,3-dioxole-co-tetrafluoroethylene) containing nanoscale fumed silica. *Macromolecules* **2003**, *36*, 8406–8414. [CrossRef]

44. Lagorsse, S.; Magalhães, F.D.; Mendes, A. Carbon molecular sieve membranes: Sorption, kinetic and structural characterization. *J. Membr. Sci.* **2004**, *241*, 275–287. [CrossRef]

membranes

MDPI

Article

Exploring the Gas-Permeation Properties of Proton-Conducting Membranes Based on Protic Imidazolium Ionic Liquids: Application in Natural Gas Processing

Parashuram Kallem [1,2,3], Christophe Charmette [2], Martin Drobek [2], Anne Julbe [2], Reyes Mallada [1,4] and Maria Pilar Pina [1,4,*

[1]   Department of Chemical & Environmental Engineering, Institute of Nanoscience of Aragon, University of Zaragoza, Edif. I+D+i, Campus Rio Ebro, C/Mariano Esquillor, 50018 Zaragoza, Spain; parshukallem@gmail.com (P.K.); rmallada@unizar.es (R.M.)

[2]   IEM (Institut Européen des Membranes), UMR 5635 (CNRS-ENSCM-UM), Université de Montpellier, CC047, Place Eugène Bataillon, 34095 Montpellier, France; Christophe.Charmette@univ-montp2.fr (C.C.); martin.drobek@univ-montp2.fr (M.D.); anne.julbe@univ-montp2.fr (A.J.)

[3]   School of Earth Sciences and Environmental Engineering, Gwangju Institute of Science and Technology (GIST), 261 Cheomdangwagi-ro, Buk-gu, Gwangju 61005, Korea

[4]   Networking Research Center on Bioengineering, Biomaterials and Nanomedicine, CIBER-BBN, 50018 Zaragoza, Spain

*   Correspondence: mapina@unizar.es; Tel.: +34-976-761155

Received: 2 August 2018; Accepted: 27 August 2018; Published: 5 September 2018

Abstract: This experimental study explores the potential of supported ionic liquid membranes (SILMs) based on protic imidazolium ionic liquids (ILs) and randomly nanoporous polybenzimidazole (PBI) supports for CH_4/N_2 separation. In particular, three classes of SILMs have been prepared by the infiltration of porous PBI membranes with different protic moieties: 1-H-3-methylimidazolium bis (trifluoromethane sulfonyl)imide; 1-H-3-vinylimidazolium bis(trifluoromethane sulfonyl)imide followed by in situ ultraviolet (UV) polymerization to poly[1-(3H-imidazolium)ethylene] bis(trifluoromethanesulfonyl)imide. The polymerization process has been monitored by Fourier transform infrared (FTIR) spectroscopy and the concentration of the protic entities in the SILMs has been evaluated by thermogravimetric analysis (TGA). Single gas permeability values of N_2 and CH_4 at 313 K, 333 K and 363 K were obtained from a series of experiments conducted in a batch gas permeance system. The results obtained were assessed in terms of the preferential cavity formation and favorable solvation of methane in the apolar domains of the protic ionic network. The most attractive behavior exhibited poly[1-(3H-imidazolium)ethylene]bis(trifluoromethanesulfonyl)imide polymeric ionic liquid (PIL) cross-linked with 1% divinylbenzene supported membranes, showing stable performance when increasing the upstream pressure. The CH_4/N_2 permselectivity value of 2.1 with CH_4 permeability of 156 Barrer at 363 K suggests that the transport mechanism of the as-prepared SILMs is solubility-dominated.

Keywords: protic imidazolium ionic liquids; CH_4 solubility; nanoporous polybenzimidazole membranes; supported ionic liquid membranes; photo-assisted polymerization; CH_4 selective membranes

1. Introduction

The demand for natural gas (NG) is growing worldwide and there is a rising need to develop methods for upgrading sub-quality gas reserves, which exist in relatively low quantities in remote zones. The global utilization of NG is above 3.1 trillion cubic meters (110 trillion standard cubic feet)

each year. NG upgrading is certainly one of the most challenging industrial applications for gas separation membranes. In fact, 14% of U.S. NG resources comprise N_2 in significant amounts and cannot be shipped to the national pipeline without preliminary treatment. Hence, removal of this N_2 could allow access to an estimated 10 trillion scf (standard cubic feet per day) additional NG in the USA alone [1–3].

So far, only a few studies on N_2 removal from methane mixtures have been published. Membrane-based N_2 separation has a promising market in small natural gas operations, where cryo-genic distillation is uneconomical. In general, glassy polymers are permeable to N_2, while the rubbery ones are to CH_4 [2]. For a gas mixture containing 10% N_2 in CH_4, a membrane with a N_2/CH_4 selectivity of at least 17 is required to achieve attractive separation in a single stage. However, the best N_2-selective membrane currently known has a selectivity of 12.5 and permeability of 0.8 Barrer [4]; i.e., far below the attractive values. Hence, this is why the CH_4-selective membranes are usually preferred. A process involving a CH_4 selective membranes process remains the most feasible. Although considerable recompression of the permeate gas is required for gas delivery to the pipeline [2], its cost does not significantly impact on the process economics [1]. For a gas mixture containing 10% N_2 in CH_4, membrane-based separation becomes cost-effective for CH_4/N_2 selectivity values above 6 [3]. However, the best CH_4-selective membrane (Polyamide-polyether copolymer-PEBAX 2533) currently known has a CH_4/N_2 selectivity of 4.2 and relatively low CH_4 permeability values, i.e., 20 Barrer.

Typically, supported ionic liquid membranes (SILMs) have been extensively studied for CO_2 separation [5–8], thanks to their good CO_2 solubility and negligible vapor pressure; although few studies have also focused on NG upgrading [9]. In general, the possible displacement of the liquid phase in SILMs is strongly diminished and more stable membranes are obtained due to both high ionic liquids (ILs) viscosity and strong capillary forces between the IL and the supporting membrane [8,10]. The most commonly used ILs are composed of imidazolium (IM) or pyridinium (Py) cations containing one or more alkyl groups, because of their low melting points and stability under a wide range of experimental conditions. Commonly used anions include halogen atoms [11], such as tetrafluoroborate $[BF_4]^-$, hexafluorophosphate $[PF_6]^-$, and bis(trifluoromethylsulfonyl)imide $[TFSI]^-$. Previous publications on SILMs confirm that the selectivity is solubility-dominated instead of diffusion-dominated [9]. The solubilities of CO_2, CH_4, C_2H_6, N_2 and O_2 in several aprotic ILs have been studied extensively [12–15]. On the contrary, the thermodynamic properties of protic ionic liquids, i.e., those comprising proton-donor and proton-acceptor centers in their molecules, have been investigated in lower extent [16–19].

Supported poly-ionic liquids (PILs) membranes based on protic imidazolium moieties have attracted great attention over the last decade as solid state flexible electrolytes because of their proton conductivity and superior thermal and chemical stability [20–24]. The main objective of this work is the exploration of the SILMs based on protic imidazolium ILs for potential CH_4 separation applications. Among the large diversity of ILs, those based on the TFSI anion with imidazolium cation typically confer high CH_4 permeability [25,26]. So far, all the reported SILMs for gas permeation studies have been prepared from aprotic ILs [7,27–29]. Unlike in the literature, our approach relies on the use of protic ILs i.e., imidazolium cation without any alkyl group at position 1 (R–N) but with acidic "H" (H–N).

Herein, we report for the first time usage of SILMs based on protic imidazolium ILs supported on/in randomly nanoporous polybenzimidazole (PBI) for gas separation of apolar compounds, i.e., CH_4 and N_2. The porous PBI employed for membrane fabrication as the mechanical support provides outstanding thermal and chemical stability [30]. In general, PBI exhibits very low gas permeability because of the carbon chain rigidity and strong intermolecular hydrogen bonding leading to dense packing structures [31,32]. More specifically, three classes of SILMs containing: (i) 1-H-3-methylimidazolium bis(trifluoromethane sulfonyl)imide (denoted as RPBI-IL); (ii) 1-H-3-vinylimidazolium bis(trifluoromethane sulfonyl)imide (denoted as RPBI-MIL); and (iii) poly[1-(3H-imidazolium)ethylene] bis(trifluoromethanesulfonyl)imide (denoted as RPBI-PIL) have been prepared. The RPBI-PIL family, obtained by the ultraviolet (UV) polymerization of the RPBI-MIL set, has been studied to improve the membrane's long-term performance [33]. Indeed, IL leaching from the pores at either high temperatures or transmembrane pressures might clearly inhibit the practical use of such membranes in gas separation processes. In addition, the polymerization phase transition from liquid to solid state effectively improves the stability of the IL-based membranes [10,33,34]. Thus, a comprehensive physicochemical and single gas permeation characterization of such SILMs has been accomplished in this work. Particular emphasis is devoted to the analysis of methane solubility and to the influence of the protic cationic moieties on the gas permeability values.

2. Methods

2.1. Chemicals

All chemical reagents and solvents were used as received: Poly[2,2-(m-phenylene)-5,5bibenzimidazole] (PBI Fumion APH Ionomer, Mw = 59,000–62,000, Fumatech), LiCl (99 wt%, Sigma-Aldrich), Poly(vinylpyrrolidone) K30 (PVP K30 Mw = 40,000, Fluka), Poly(vinylpyrrolidone) K90 (PVP K90 Mw = 360,000, Fluka), 1-H-3-methylimidazolium bis(trifluoromethane sulfonyl)imide (99.5 wt%, Solvionic), 1-H-3-vinylimidazolium bis(trifluoromethane sulfonyl)imide (99.5 wt%, Solvionic), Divinylbenzene (80.0 wt%, Sigma-Aldrich), 2-hidroxy-2-methylpropiophenone (97.0 wt%, Sigma-Aldrich), N-methyl-2pyrrolidone (NMP anhydrous, 99.5 wt%, Sigma-Aldrich).

2.2. Polymer Solution Preparation

PBI was used as a polymer for the fabrication of membrane support. Polymer solutions were prepared according to a recipe previously developed in our group [21]. 11.5 g of PBI powder, 1.5 g of LiCl, 1.5 g of PVP K30, 1.5 g of PVP K90 and 84 g NMP were mixed at 448 K for 24 h to obtain 16 wt% of solids in homogeneous polymer solution. The polymer solution was then degassed under moderate vacuum for two hours to ensure that all air bubbles were removed from the solution. Addition of PVP controls macrovoids formation upon phase separation process and LiCl stabilizes the polymer solution. Before use, the PBI solution was filtered by pressurized air through metal filter (25 m in pore size) to remove insoluble solids from the starting PBI powder.

2.3. Preparation of the Randomly Porous Polybenzimidazole (RPBI) Supports by Phase Inversion

A schematic overview of the phase inversion process is depicted in Figure S2 (supplementary materials). The polymer solution consisting of PBI, PVP, LiCl and NMP was poured onto a clean glass plate (Figure S2A) and cast using a casting knife with a thickness of 0.25 mm. After casting, the glass plate with deposited polymer layer was immersed in a coagulation bath (Figure S2B) containing a mixture of NMP/water (50/50) for 30 min at room temperature (RT ~298 K). Then the plate was transferred into a non-solvent bath (pure water) at room temperature (RT) to wash out any NMP traces (Figure S2C), the exchange of solvent by water was effective after 30 min at RT. The polymer film was then peeled off from the plate. Subsequently solidified RPBI support was immersed in ethanol for 30 min, followed by an immersion in hexane for 30 min, to ensure complete water removal. Finally, remaining volatiles were evacuated at 423 K in an oven. For this thermal treatment, the polymer film was sandwiched between two glass plates.

2.4. Fabrication of Supported Ionic Liquid Membranes (SILMs)

The RPBI supports were infiltrated with ionic liquids using a protocol previously developed by the authors [21]. A schematic illustration of the infiltration protocol is shown in Figure 1. Firstly, the RPBI support was dried at 393 K under 100 mbar of vacuum to remove any water and organics. Three types of SILMs have been fabricated:

(i) RPBI-IL: the protic ionic liquid (IL) 1-H-3-methylimidazolium bis(trifluoromethane sulfonyl)imide (H-MIM TFSI) was heated up to 328 K to melt the salt. Subsequently, the RPBI support was placed under vacuum for 1h to remove air from the pores and guarantee an efficient and uniform (in)filtration of the IL through the RPBI support. The infiltration process was conducted by pouring the IL on the RPBI support surface at 443 K and applying 160 mbar vacuum. After the filtration step, the membrane was removed from the filter holder and the excess of IL on the membrane surface was wiped off with a tissue.

(ii) RPBI-MIL: the monomeric ionic liquid (MIL) 1-vinyl-3H-imidazolium bis(trifluoromethane sulfonyl)imide (H-VIM TFSI) was heated up to 323 K to melt the salt, and the above (in)filtration protocol was applied.

(iii) RPBI-PIL: the monomeric ionic liquid (MIL) 1-vinyl-3H-limidazolium bis(trifluoromethane sulfonyl)imide was heated up to 323 K to melt the salt; afterwards 1 mol% (referred to the MIL) of divinylbenzene (crosslinker-CL) was added and the mixture was thoroughly mixed. Subsequently, the above (in)filtration protocol was applied. Finally, a photo initiator (2-hydroxy-2-methylpropiophenone) was added on the membrane top-surface to initiate photo-polymerization. In order to obtain the crosslinked RPBI/PIL composite membranes, each side of the membrane surface was exposed for 2 h under a 365 nm UV lamp (Vilber Lourmat, Collégien, France) with intensity of 2.4 mW cm^{-2}). After IL polymerization, the membrane surface was gently wiped from any residuals with a lab paper.

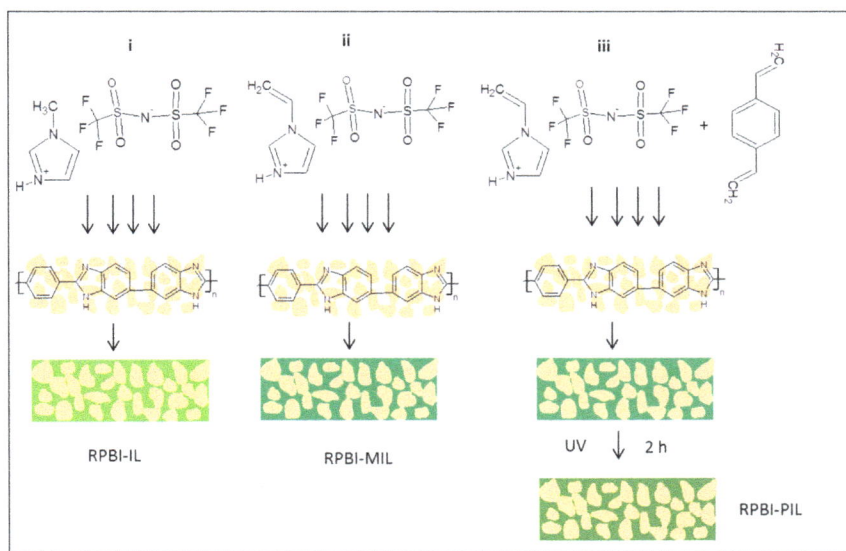

Figure 1. Schematic illustration of polybenzimidazole (PBI) support pore filling and chemical structures of the used ionic liquids: (i) 1-H-3-methylimidazolium bis(trifluoromethane sulfonyl)imide (H-MIM TFSI); (ii) 1-vinyl-3H-imidazolium bis(trifluoromethane sulfonyl)imide (H-VIM TFSI); (iii) H-VIM TFSI with divinylbenzene followed by polymerization with UV light.

2.5. Characterization Methods

Porosity: The porosity of the as prepared RPBI support was determined by using a helium displacement pycnometer (Micromeritics AccuPyc 1330, Micromeritics Instrument Corp., Norcross, GA, USA) equipped with 1 cm^3 sample module. The reported porosity values were obtained for RPBI supports with more than 50 cm^2 surface area. Porosity was calculated using following equation:

$$\text{Porosity} \ (\varnothing) = \frac{\text{Vbulk} - \text{Vskeleton}}{\text{Vbulk}} \times 100\% \tag{1}$$

where Vbulk is directly estimated from surface area and thickness of the RPBI sample and Vskeleton is obtained from the instrument.

For all samples, measurement reproducibility was typically within ±0.01% of the nominal porosity value.

Scanning electron microscopy (SEM) characterization: the morphology, thickness, porous structure and pore size of the as-prepared RPBI supports were observed by scanning electron microscopy (SEM) (FEI INSPECT 50), acceleration voltage 15 keV. Prior to observation, the samples were coated with a Pd layer of ca. 2 nm by sputtering (LEICA EM ACE200).

Transmission electron microscopy (TEM): membranes were embedded in epoxy resin, and ultrathin slices (about 50 nm thick) were cut with an ultramicrotome (Leica EM UC7) at room temperature. These slices were placed on TEM copper grids with carbon film, and analyzed by TEM in a Tecnai T20 (FEI Company), at a working voltage of 200 KV. TEM bright field images were acquired with a side-mounted Veleta CCD Camera.

Atomic force microscopy (AFM): AFM measurements have been carried out by tapping mode using NSG30 ND-MDT tip (Multimode 8 system, Veeco/Bruker) with force constant around 22–100 N/m. Roughness average (Ra) and root mean square (RMS) values are both representations of surface roughness, although calculated differently from microscopic peaks and valleys on the surface using the following equations:

$$Ra = \frac{1}{N} \sum_{i=1}^{n} [yi] \tag{2}$$

$$RMS = \sqrt{\frac{1}{N} \sum_{i=1}^{n} yi} \tag{3}$$

The roughness profile contains N ordered, equally spaced points along the trace, and yi is the vertical distance, expressed in nm, from the mean line to the ith data point.

Infrared spectra measurements: attenuated total reflection–Fourier transform infrared (ATR-FTIR) analyses (Bruker VERTEX 70 equipped with Golden Gate ATR from 4000 to 600 cm^{-1}, 256 scans and resolution of 4 cm^{-1}) were performed at room temperature to assess about the photo-polymerization evolution in RPBI/PIL SILMs, and to investigate any possible interactions between the benzimidazole from the RPBI support and the poly[1-(3H-imidazolium)ethylene] bis(trifluoromethane sulfonyl)imide.

Thermogravimetric studies: thermogravimetric analyses (TGA) were carried out using a Q500 IR TA instrument to evaluate the composition and thermal behavior of the as-prepared SILMs. Studies were conducted using 4–5 mg samples, in the temperature range from room temperature up to 1173 K at a controlled heating rate of 2 K/min under an inert atmosphere (N_2).

Methane solubility in the protic ionic liquid: the CH_4 gas solubility in the H-VIM TFSI was calculated by using the experimental vapour pressure equilibrium. The vapor pressure of the protic ionic liquid mixture with methane was measured at 333 K at five compositions (from 0.0056 to 0.0165 methane molar fraction) in the experimental set-up described by Coronas et al. [35] using a static isochoric method.

Single gas permeation experiments: single gas permeation measurements through the membranes were carried out by using the constant-volume and variable-pressure technique at controlled temperature, as described in the standard ASTM D 1434-82 protocol (procedure V). A schematic of the experimental set-up (home-made) is shown in Figure S3. The two compartments of the permeation cell are separated by the tested membrane. The permeability was obtained by measuring the pressure increase in the downstream compartment (with a constant volume of 5.25 10^{-5} m^3) and using different MKS Baratron pressure transducers (range from 0.0 to 1 × 10^5 Pa). The membrane and downstream cell walls were initially outgassed in situ during 15 h at high vacuum using a turbomolecular pump (Leybold, Turbovac 50). Permeability values were measured in the temperature range from 313 K to 363 K, using classically up to 1.5 × 10^5 Pa of upstream pressure gauge (unless otherwise indicated). The pressure increase in the downstream compartment was continuously measured during 4 h. For each temperature change, the whole set-up was stabilized during at least 12 h. For a given temperature, the order of gas permeance measurements was as follows: N_2, CH_4. Between each measurement, both the membrane and the cell were outgassed in situ during 12 h under high vacuum.

Both N_2 and CH_4 were provided by Linde Gas as single gases with 99.95% purity, and were used without any further purification. A complete description of the experimental system and measurement protocol was reported elsewhere [36]. For permeability calculations, a mathematical treatment relevant for thin films and based on the second Fick's law was used:

$$P = \frac{V\,L}{A\,R\,T\,P_1}\left(\frac{dP_2}{dt}\right) \tag{4}$$

where P (mol m^{-1} s^{-1} Pa^{-1}) is the gas permeability; V (m^3) is the volume of the downstream compartment; L (m) is the membrane thickness; A (m^2) is the membrane surface area; R is the universal gas constant (Pa m^3 mol^{-1} K^{-1}); T is the permeation temperature (K); P_1 (Pa) is the applied feed side pressure; and P_2 (Pa) is the recorded pressure at the permeate side.

3. Results and Discussion

3.1. Fabrication of the Randomly Porous Polybenzimidazole (RPBI) Supports: Morphological Characterization

The RPBI supports, 120 to 175 μm thick, were prepared successfully by a phase separation method already reported in our previous work [21]. The porosity measured by pycnometry was 63.7 ± 2.7%. SEM pictures of the prepared RPBI supports are shown in Figure 2, where the analysis of airside, glass side and cross-section are displayed. Pore sizes in the range 50–250 nm were measured on the air side and 30–50 nm on the glass side. The cross-section view reveals a sponge-like structure.

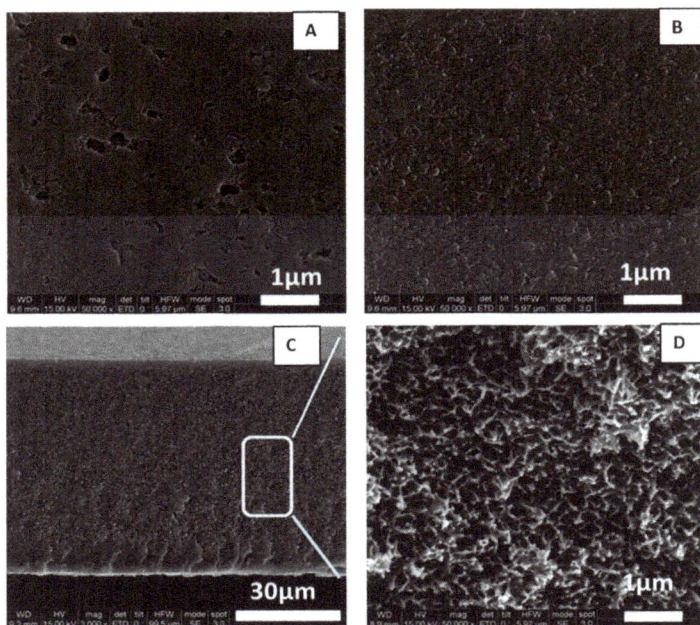

Figure 2. Scanning electron microscope (SEM) observation of a RPBI support prepared by phase separation method from 16 wt% of solid in the polymer solution: (**A**) air (top) side; (**B**) glass (bottom) side; (**C**) cross-section; (**D**) detail of cross-section area.

To better understand the pore connectivity, the microstructure of RPBI support was observed by TEM. A typical image of the cross-section is shown in Figure 3A. The clear regions correspond to the pores and interconnections between random pores can be observed over the whole membrane thickness. In order to examine the surface roughness of the RPBI support, AFM surface images of both glass and air side (bottom and top side, respectively) were analyzed (Figure 3B,C). The minor changes in roughness parameters (roughness average and root mean squared roughness values reported on the AFM images) from top to bottom are attributed to the change in the size of interconnected open pores.

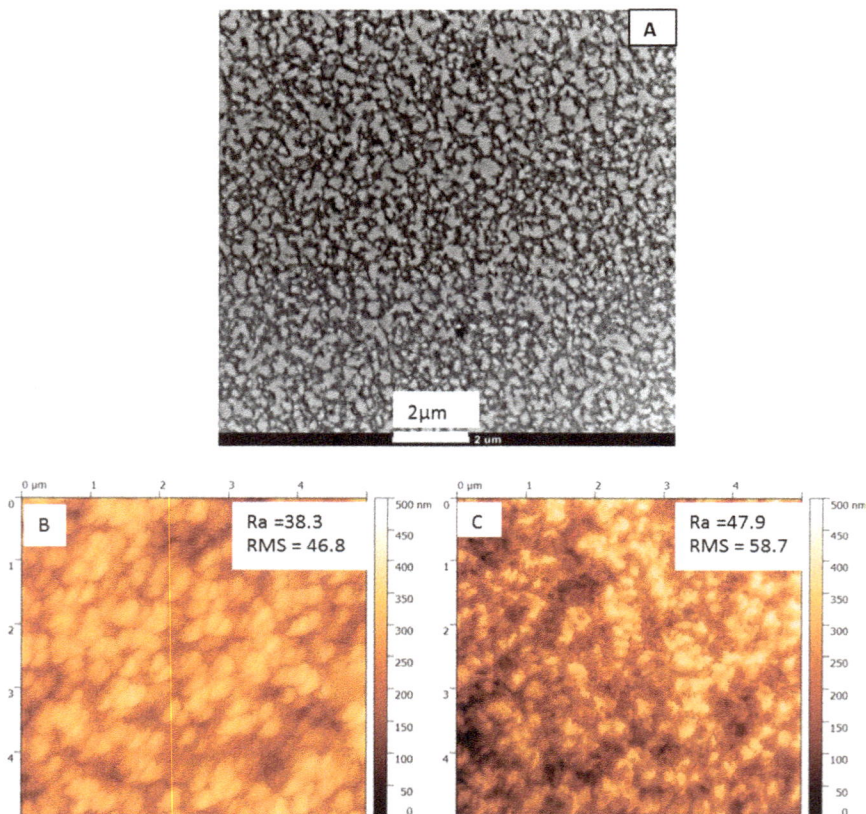

Figure 3. (**A**) Transmission electron microscope (TEM) observation of randomly porous PBI (RPBI) support and AFM surface images of (**B**) RPBI-glass (bottom) side and RPBI-air (top) side (**C**). The values of roughness average (Ra) and root mean squared (RMS) roughness, expressed in nm, are reported on the atomic force microscope (AFM) images.

3.2. Fabrication of SILMs Based on Protic Imidazolium Ionic Liquids: Physico-Chemical Characterization

SILMs were prepared by infiltration of RPBI support with the protic ionic liquids, 1-H-3-methylimidazolium bis(trifluoromethane sulfonyl)imide monomeric ionic liquid (H-MIM TFSI) and 1-H-3-vinylimidazolium bis(trifluoromethane sulfonyl)imide (H-VIM TFSI), respectively, as described in the experimental section. Due to the viscosity of H-MIM TFSI (i.e., 80 cP at 298 K) [22] and H-VIM TFSI (14.3 cP at 323 K), the use of both vacuum and high temperature was required to ensure efficient IL embedding within the pores of the RPBI support. A schematic illustration of the infiltration protocol is shown in Figure 1A.

FTIR analyses were used to evidence the successful polymerization of vinyl-polymerizable groups. Accordingly, ATR-FTIR spectra of the composite membranes before (RPBI-MIL), and after 2 h UV irradiation (RPBI-PIL) are compared in Figure 4. The intense absorption bands in the range 1400–1000 cm^{-1}, observed for both RPBI-MIL and RPBI-PIL membranes, are characteristic of the –SO$_2$– and –SNS– vibrational modes of the bis(trifluoromethanesulfonyl)imide [TFSI] anion [37]. Two characteristic infrared absorbance bands in RPBI-MIL were selected to examine the disappearance of the vinyl-monomer: 1665–1630 cm^{-1} (stretching vibration in –CH=CH$_2$) and 995–920 cm^{-1} (out of plane bending of –CH=CH$_2$ groups). The disappearance of these characteristic peaks in RPBI-PIL upon 2 h UV light exposure confirmed a successful polymerization, above 97%, as already demonstrated in our previous studies [20,21,23,24]. Figure S1 (supplementary material) shows photos of the prepared SILMs, as free-standing films. The SILMs based on IL (i.e., RPBI-IL) were extremely brittle and hard to handle, due to the IL crystallinity at room temperature (melting point ~328 K) whereas the RPBI-MIL was only slightly brittle when handling (melting point ~313–318 K). Contrary, the SILMs based on polymeric IL (i.e., RPBI-PIL) were very easy to manipulate.

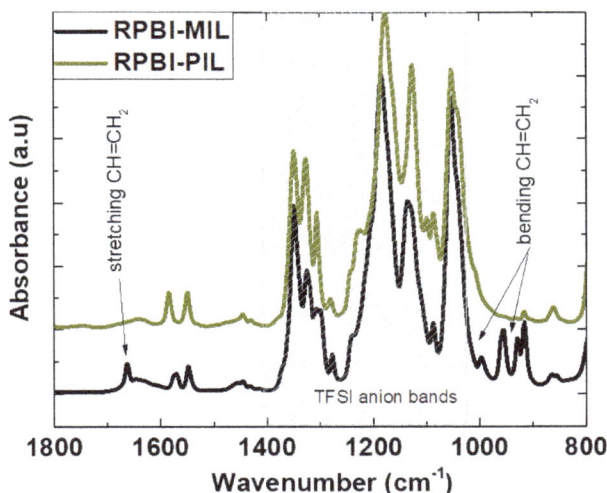

Figure 4. Attenuated total reflection–Fourier transform infrared (ATR-FTIR) spectra of the resulting SILM membrane before and after UV irradiation.

Table 1 summarizes the characteristics of the as prepared SILMs. The experimental IL/MIL/PIL loadings calculated from simple weight increase measurements reasonably match those evaluated from TGA (accounting from the registered weight loss within the 473–778 K temperature range), but overpass theoretical values due to the difficulty in wiping the excess of IL/MIL/PIL from the membrane surface. Therefore, the infiltration process herein performed ensures complete pore filling of the PBI supports.

Table 1. Main characteristics of SILMs based on protic imidazolium moieties specifically prepared for this work.

SILM	Ionic liquid (IL) Loading (wt%)		
	Theoretical [1]	Experimental [2]	TGA
RPBI-IL	73.5	82.4	70.4
RPBI-MIL	77.0	86.0	83.1
RPBI-PIL	82.5	86.5	78.6

[1] Theoretical calculations based on both IL/MIL/PIL density and membrane porosity; [2] Experimental estimation from weight measurements.

Figure 5 shows the TGA and derived DTG curves of all the prepared SILMs. A one stage thermal decomposition process (663 K) was observed for both RPBI-IL and RPBI-MIL samples due to the decomposition of H-MIM TFSI and H-VIM TFSI, respectively. Whereas in the case of RPBI-PIL membrane a two-stage decomposition was observed: the first weight loss corresponds to PIL decomposition and the shoulder at ~753 K is attributed to interactions between PIL and RPBI support [21]. Very low weight loss (0.7–1.5%) were measured for all SILMs within the 423–473 K temperature range, suggesting a good thermal stability of SILMs up to 583 K under N_2 atmosphere. The measured Young's modulus and tensile strength values of RPBI-PIL were 0.2 GPa and 1.3 MPa, respectively (refer to our previous work [23] for more details). The RPBI-IL and RPBI-MIL samples were not considered for mechanical tests due to handling difficulties at room temperature.

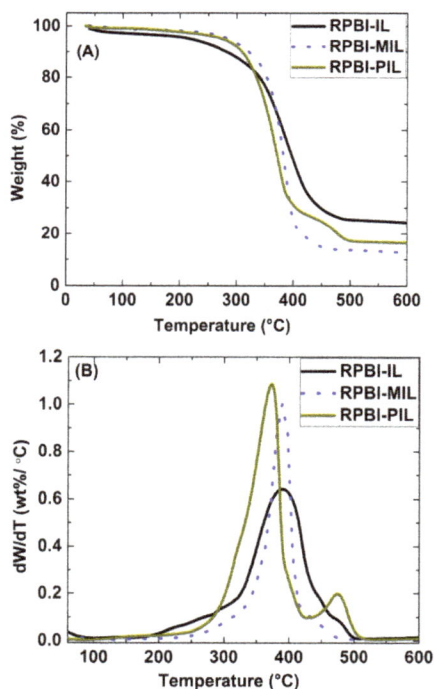

Figure 5. (**A**) Thermogravimetric analysis (TGA) curves and (**B**) derived differential (DTG) curves of the prepared supported ionic liquid membranes.

3.3. Permeation Properties of the SILMs Based on Protic Imidazolium Ionic Liquids

To the best of our knowledge, the permeation properties of SILMs based on protic imidazolium ionic liquids have been scarcely investigated in the literature. In this work, the single gas permeances of N_2, CH_4 were measured in order to evaluate the membrane permselectivity (PermSel CH_4/N_2) for CH_4/N_2 separation which is an important parameter for possible upgrading of natural gas.

In parallel, solubility data of gases in ionic liquids are required for designing the separation processes and provide the basis for tuning the ionic liquids properties. The potential of ionic liquids for the separation of CH_4/N_2 gas mixture can be evaluated by the ideal selectivity (Ideal Sel CH_4/N_2) which is defined by the ratio of Henry's constant values (H N_2/H CH_4).

Table 2 compares the Henry's constant values for CH_4 and N_2 at different temperatures in the range 298–343 K, in common aprotic ionic liquids based on methylimidazolium cations and [TFSI] anion. CH_4 is the most soluble (lowest Henry's constant), while the solubility of N_2 is lower (higher Henry's constants) for all the tested conditions. Regular solution theory has been extensively used as a method to model the behavior of gases in aprotic ILs. The widely investigated CO_2 + IL system could be accurately modeled as a function of the sorbent molar volume, with smaller molar volumes and lower temperatures yielding both increasingly higher CO_2 solubilities and ideal CO_2/gas selectivities.

Table 2. Values of Henry's law constant for N_2 and CH_4 in different aprotic ILs and derived calculated ideal selectivities.

Aprotic Ionic Liquids	T (K)	H N_2 (atm)	H CH_4 (atm)	Ideal Sel CH_4/N_2	Ref.
1-hexyl-3-methylimidazolium bis(trifluoromethanesulfonyl)imide	298	1000 ± 8	350 ± 1	2.8	[14]
	313	830 ± 6	350 ± 2	2.4	
	328	720 ± 11	340 ± 4	2.1	
	343	660 ± 12	340 ± 0.4	1.9	
1-butyl-3-methylimidazolium bis(trifluoromethanesulfonyl)imide	333	970 ± 30	420 ± 10	2.3	[38]
1-ethyl-3-methylimidazolium bis(trifluoromethanesulfonyl)imide	298	1400 ± 17	580 ± 4	2.9	[14]
	313	1200 ± 27	560 ± 3	2.1	
	328	1000 ± 19	540 ± 1	1.85	
	343	910 ± 0.3	530 ± 0.4	1.7	

Table 2 displays the solubility selectivity trend for CH_4 and N_2 pairs as a function of temperature. The same solubility selectivity trends exist for all the aprotic ionic liquids tested. With increasing temperature, the solubility selectivity slightly decreases, i.e. from 2.9 at 298 K to 1.7 at 343 K. This behavior was expected when considering the observed evolution trend of solubility vs. temperature. Unlike the CO_2 + ILs equilibria behaviors, the solubility of N_2 increases (decreasing Henry's constant) when temperature increases for all the aprotic ionic liquids. On the other hand, the CH_4 solubility remains almost constant, indicating that the change in partial molar enthalpy and entropy of the system must be zero (based on thermodynamic equations). Additionally, the Henry´s constant value for CH_4 decreases when the molar volume of the aprotic IL increases, i.e., from 580 atm to 350 atm at 298 K for 1-ethyl-3-methylimidazolium bis(trifluoromethanesulfonyl)imide and -hexyl-3-methylimidazolium bis(trifluoromethanesulfonyl)imide, respectively. When comparing the ideal selectivity values obtained from solubility measurements (Sel CH_4/N_2) with those corresponding to membrane permselectivity (Perm Sel CH_4/N_2), a solubility dominated transport was confirmed for aprotic ionic liquids [9].

Table 3 compares the Henry´s constant values for CH_4 and N_2 in the temperature range 303–333 K, in protic ionic liquids, i.e., those containing proton-donor and proton-acceptor centers in their molecules. Protic ionic liquids are described as structurally heterogeneous compounds consisting of both polar and apolar domains. Charged and uncharged groups tend to segregate resulting in sponge-like nanostructures. In general, the thermodynamic properties of ionic liquids with dissociable protons are significantly less investigated than those of aprotic analogues. An inert-gas stripping method has been described in the literature [16] for measuring solubilities of moderately and sparingly soluble gases, i.e., N_2, O_2, air, C_2H_4, C_2H_6 in low viscosity protic ionic liquids such as 1-butyl, 3-H-imidazolium acetate. In protic ionic liquids, CH_4 is the most soluble (lowest Henry's constant) while the solubility of N_2 lower (higher Henry's constant) than in their aprotic imidazolium counterparts. The CH_4 Henry's constant for the 1-H-3-vinylimidazolium bis(trifluoromethanesulfonyl)imide ionic liquid used in this work is 172 ± 16 atm at 333 K, i.e., three fold lower than the values measured for the aprotic 1-ethyl-3-methylimidazolium bis(trifluoromethanesulfonyl)imide. At a first glance, the intermolecular hydrogen bonds in protic molecular solvents seem to yield a significant drop of solubility for apolar species. However, here we do observe a tendency of higher CH_4 solubility in comparison with similar aprotic ionic liquids. Sedov et al. [17] explain this behaviour by a preferential cavity formation and favorable solvation of hydrocarbons in the apolar domain of nanostructured protic ionic liquids. Consequently, the higher CH_4 permeability values measured for SILM$_S$ prepared from protic ionic liquids would be expected.

Table 3. Values of Henry's law constant for N_2 and CH_4 in different imidazolium based protic ILs and derived ideal selectivity values.

Protic Ionic Liquids	T (K)	H N_2 (atm)	H CH_4 (atm)	Ideal Sel CH_4/N_2	Ref.
1-butyl-3-H-imidazolium acetate	308	1840 ± 147	90 ± 4.5 * 85 ± 3.4 **	20.4 * 21.6 **	[16]
1-H-3-vinylimidazolium bis(trifluoromethanesulfonyl)imide	333	n.a.	172 ± 16	n.a.	This work

* evaluated for C_2H_6; ** evaluated for C_2H_4.

In this work, the single gas permeability results were assumed to reflect gas transport through protic ionic liquid moieties while the contribution of the parallel RPBI transport pathway was considered as negligible due to the extremely low permeabilities of gases in dense PBI [39], i.e., 0.009 Barrer for CH_4. The influence of temperature on both gas permeability (Figure 6) and CH_4/N_2 permselectivity (Figure 7) was studied for the three different SILMs in the temperatures range 313–363 K and measured in the initial pressure range of 1.5 barg. Since H-MIM TFSI is a crystalline solid at room temperature and its melting point is ~328 K [40], the experiments with RPBI-IL were carried out at 333 K and above.

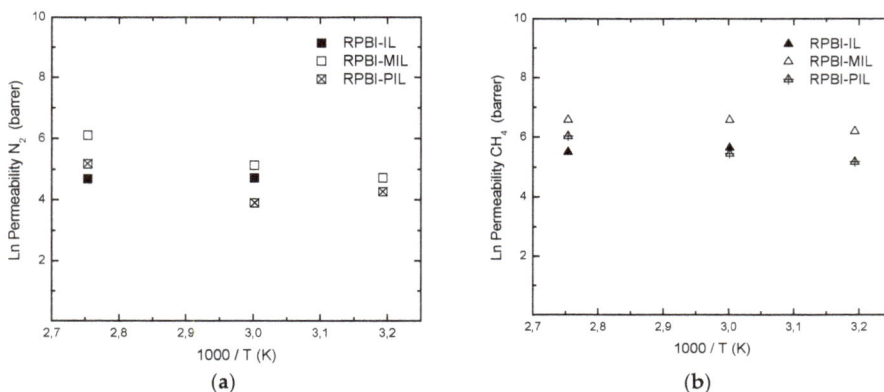

Figure 6. Influence of temperature on single gas permeability values for: N_2 (**a**), CH_4 (**b**).

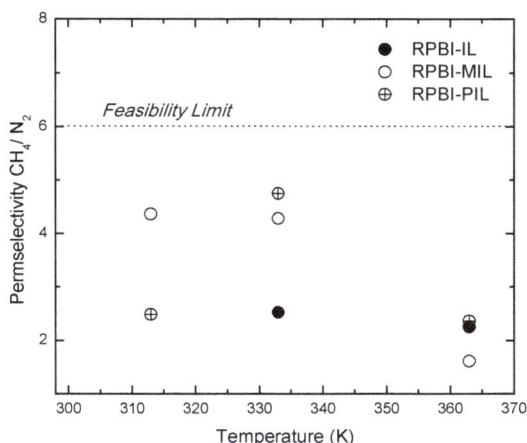

Figure 7. Influence of temperature on CH_4/N_2 permselectivity values.

The measured permeability values of the prepared SILMs are in the range 49–178 Barrer for N_2 and 178–725 Barrer for CH_4. In the tested temperature range, i.e., 333–363 K, the CH_4 permeability of RPBI-MIL membranes is always higher than for RPBI-IL membranes. Hence, the effect of vinyl substitution on imidazolium group seems to increase the CH_4 solubility. According to Scovazzo et al. [9], in addition to the consideration of IL viscosity and molar volume, the IL ability to accept hydrogen to form a hydrogen bond contributes to a better correlation of the permeance trends for N_2, CH_4 and C_xH_y through SILMs. In our previous work [20] on the use of protic ionic liquids for the preparation of all solid state ion conductive films, the proton transport properties of HVIM TFSI were found to be superior to those of the HMIM TFSI counterpart. Hence, proton conduction properties seem to be in line with observed CH_4 permeation values.

All solid-state gas permeable membranes, denoted as RPBI-PIL, were also prepared by UV photo-assisted polymerization of supported HVIM TFSI membrane to provide SILMs with adequate physical stability for gas separation applications involving moderate to high trans-membrane pressures. As expected, the cationic moieties polymerization strongly impacts the membrane permeation properties. In fact, gas permeability values of RPBI-PIL membranes were three times lower than those measured for RPBI-MIL membranes (Table 4). Above all, when compared with the RPBI-IL counterparts, the CH_4 permeance through RPBI-PIL resembles the same at the expense of a remarkable improvement of endurance properties.

Table 4. Single gas permeability values measured for the SILMs prepared in this work and derived permselectivity values.

Ionic Liquid	Support	Temperature (K)	N_2 (Barrer)	CH_4 (Barrer)	PermSel CH_4/N_2
1-H-3-methylimidazolium bis(trifluoromethane sulfonyl)imide	RPBI	333	112	285	2.5
1-H-3-vinylimidazolium bis(trifluoromethanesulfonyl)imide	RPBI	333	169	725	4.3
poly [1-(3H-imidazolium) ethylene] bis (trifluoromethanesulfonyl)imide	RPBI	333	50	235	4.7
1-ethyl-3-methylimidazolium bis(trifluoromethanesulfonyl)imide	PVDF *	303	17	32	1.9

* Data from Reference [9]: commercial PVDF 125 μm thick, 70% porosity, 0.1 m pore diameter.

In this study, the N_2 permeability values tend to increase moderately within the tested temperature window; whereas CH_4 is less temperature dependent (Figure 6), which is in a good agreement with the gas solubility data reported in Table 2.

The calculated CH_4/N_2 PermSel values, corresponding to the ratio of single gas permeabilities, are reported in Table 4. As observed in Figure 7, the SILMs developed in this work exhibit relatively high PermSel values, up to 4.7 for RPBI-PIL at 333 K. Although this value is below 6, i.e., the target estimated by Baker [3] for cost-effective NG processing with membranes, the methane permeability through RPBI-PIL remains always above 60 Barrer for the tested temperature window.

The key difference between IL-based and polymer-based membranes is the impact of gas diffusivity on membrane selectivity. In IL-based membranes, the gas diffusivity selectivity is constant for a given gas pair, whereas the solubility selectivity controls membrane selectivity [21]. In most polymeric membranes, the opposite behavior is observed: solubility selectivity is usually constant for gas pairs and it is the diffusivity selectivity which determines the membrane selectivity [26,41]. The CH_4/N_2 permselectivity values for all the herein studied SILMs are plotted in Figure 8 as a function of CH_4 permeability.

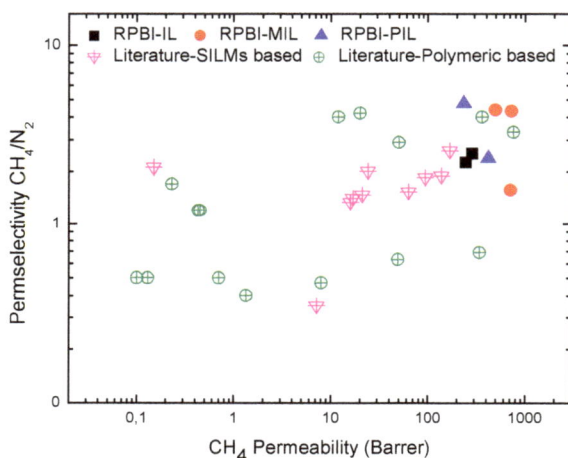

Figure 8. Comparison of permselectivity vs. permeability values for the membranes prepared in this work and for series of SILMs and polymer membranes reported in the literature.

It is evident that the performance of the SILMs developed in this work are highly promising in comparison with literature data for either polymeric membranes (adapted from both Lokhandwala et al. [1] and Scholes et al. [2]) or other SILMs (based on aprotic imidazolium cation and TFSI anion on/in different supports) [7,25–29,42,43].

Among all the prepared SILMs, the RPBI-PIL family stands for the most adequate in terms of CH_4/N_2 transport properties. In a step further, these membranes were subjected to gas permeance experiments using an initial up-stream pressure up to 4×10^5 Pa for durability evaluation. Figure 9 summarizes the results obtained for both single gases at 313 K and 363 K, respectively. A slight decline in the measured CH_4 and N_2 permeability values was observed when the initial upstream pressure increased from 1.5 barg to 2.5 barg at 363 K. Apart from this observation, the permeation properties remain constant whatever the pressure applied in the feed side: 32 to 72 Barrer for N_2 and 61 to 156 Barrer for CH_4 at 313 K and 363 K, respectively. These results confirm the expected endurance provided by the polymerization of the cationic moieties in the RPBI support.

The polymerized IL, nanoconfined in the RPBI support, have been shown to provide stable performance with both relatively high CH_4 permeability (>60 Barrer) and stable CH_4/N_2 permselectivity (in the range 2.0–4.7) up to 363 K and 4.0 barg, and attractive performance is also expected for the separation of gas mixtures with the RPBI-PIL membrane family upon long-term operation. This will be the subject of our future investigations, focusing in more details on the gas permeation and separation measurements for a long period of time.

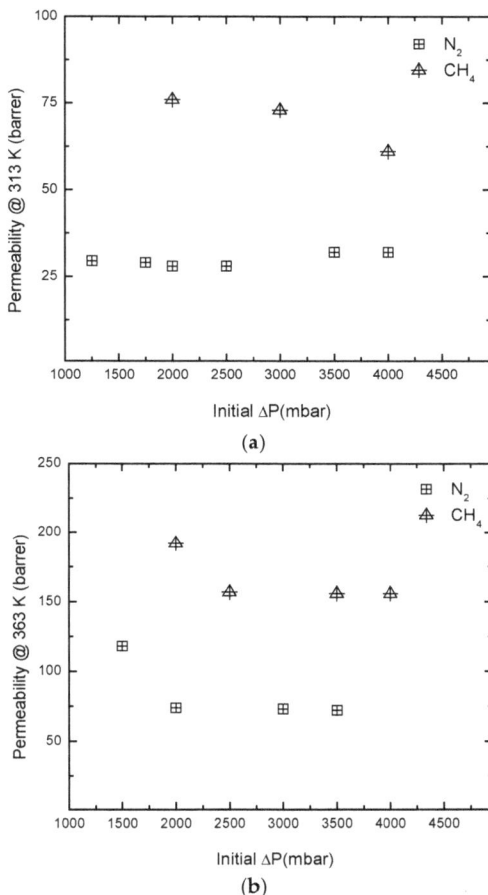

Figure 9. Influence of initial feed side pressure on single gas permeability values as a function of temperature: 313 K (**a**) and 363 K (**b**) for RPBI-PIL membranes.

4. Conclusions

In this work we presented for the first time an experimental study of N_2 and CH_4 permeation properties of supported ionic liquid membranes (SILMs) based on protic imidazolium [TFSI] ionic liquids supported in randomly nanoporous PBI (RPBI). So far, only limited studies can be found in the literature on the separation performance of bulky protic ionic liquids focusing essentially on the evaluation of the ideal CH_4/N_2 selectivity calculated from Henry's constant values. Unexpectedly, the CH_4 solubility in the 1-H-3-vinylimidazolium bis(trifluoromethanesulfonyl)imide ionic liquid used in this work is three times higher than the values measured for its similar aprotic counterparts.

This observation is attributed to the favorable solvation of hydrocarbons in the apolar domains of nanostructured protic ionic liquids.

The measured permeability values of the prepared SILMs based on 1-H-3-methylimidazolium [TFSI], 1-H-3-vinylimidazolium [TFSI] and poly[1-(3H-imidazolium)ethylene] [TFSI] were found to be in the range 49–178 Barrer for N_2 and 178–725 Barrer for CH_4 at temperatures varying from 313 to 363 K. Among the studied SILMs, those based on poly[1-(3H-imidazolium)ethylene] [TFSI] are clearly superior with CH_4/N_2 permeation properties comparable or higher than the state of the art membranes, i.e., a CH_4/N_2 permselectivity of 4.7 with a CH_4 permeability reaching 235 Barrer at 333 K. The membrane permeability is above the target and particularly attractive for industrial applications.

In addition, such solid-state gas selective poly-ionic liquid-based membranes exhibit stable performance at moderate trans-membrane pressures, i.e., 4 barg, thanks to the in situ polymerization and confinement of the cationic moieties within the pores of the RPBI support. This work is a strong motivation for future investigations of poly[1-(3H-imidazolium)ethylene] [TFSI] supported membranes in a long-term performance operation with gas mixtures relevant for natural gas upgrading.

Supplementary Materials: The following are available online at http://www.mdpi.com/2077-0375/8/3/75/s1, Figure S1: Photos of the prepared SILMs: A) IL-based SILM (RPBI-IL); B) MIL-based SILM (RPBI-MIL); C) PIL-based SILM (RPBI-PIL), Figure S2: Schematic of the phase inversion steps: (A) polymer solution casting on clean glass plate; (B) System immersed in a coagulation bath with solvent mixture 50:50% of NMP: water; (C) Glass plate with the formed PBI support immersed in pure water, Figure S3: Schematic of the lab-scale experimental set-up used for single gas permeation measurements.

Author Contributions: P.K., A.J., R.M. and M.P. conceived and designed the experiments. P.K. performed the materials synthesis and carried out the experiments. C.C. carried out the gas permeation experiments. P.K., R.M. and M.P. analysed the data and interpreted the results. A.J., M.D. contributed to discussions. P.K. and M.P. drafted the manuscript. All authors reviewed the manuscript.

Funding: This research was funded by Government of Aragon and the Education, Audiovisual, and Culture Executive Agency (EU-EACEA) within the EUDIME "Erasmus Mundus Doctorate in Membrane Engineering" program (FPA 2011-0014, SGA 2012-1719, http://eudime.unical.it).

Acknowledgments: The authors would like to acknowledge the LMA-INA for offering access to their instruments and expertise.

Conflicts of Interest: The authors declare no conflict of interest.

References

1. Lokhandwala, K.A.; Pinnau, I.; He, Z.; Amo, K.D.; DaCosta, A.R.; Wijmans, J.G.; Baker, R.W. Membrane separation of nitrogen from natural gas: A case study from membrane synthesis to commercial deployment. *J. Membr. Sci.* **2010**, *346*, 270–279. [CrossRef]

2. Scholes, C.A.; Stevens, G.W.; Kentish, S.E. Membrane gas separation applications in natural gas processing. *Fuel* **2012**, *96*, 15–28. [CrossRef]

3. Baker, R.W.; Lokhandwala, K. Natural Gas Processing with Membranes: An Overview. *Ind. Eng. Chem. Res.* **2008**, *47*, 2109–2121. [CrossRef]

4. Ohs, B.; Lohaus, J.; Wessling, M. Optimization of membrane based nitrogen removal from natural gas. *J. Membr. Sci.* **2016**, *498*, 291–301. [CrossRef]

5. Shimoyama, Y.; Komuro, S.; Jindaratsamee, P. Permeability of CO_2 through ionic liquid membranes with water vapour at feed and permeate streams. *J. Chem. Thermodyn.* **2014**, *69*, 179–185. [CrossRef]

6. Close, J.J.; Farmer, K.; Moganty, S.S.; Baltus, R.E. CO_2/N_2 separations using nanoporous alumina-supported ionic liquid membranes: Effect of the support on separation performance. *J. Memb. Sci.* **2012**, *390–391*, 201–210. [CrossRef]

7. Bara, J.E.; Gabriel, C.J.; Carlisle, T.K.; Camper, D.E.; Finotello, A.; Gin, D.L.; Noble, R.D. Gas separations in fluoroalkyl-functionalized room-temperature ionic liquids using supported liquid membranes. *Chem. Eng. J.* **2009**, *147*, 43–50. [CrossRef]

8. Dai, Z.; Noble, R.D.; Gin, D.L.; Zhang, X.; Deng, L. Combination of ionic liquids with membrane technology: A new approach for CO_2 separation. *J. Membr. Sci.* **2016**, *497*, 1–20. [CrossRef]

9. Khakpay, A.; Scovazzo, P. Reverse-selective behavior of room temperature ionic liquid based membranes for natural gas processing. *J. Membr. Sci.* **2018**, *545*, 204–212. [CrossRef]

10. Tome, L.C.; Marrucho, I.M. Ionic liquid-based materials: A platform to design engineered CO_2 separation membranes. *Chem. Soc. Rev.* **2016**, *45*, 2785–2824. [CrossRef] [PubMed]

11. Liang, L.; Gan, Q.; Nancarrow, P. Composite ionic liquid and polymer membranes for gas separation at elevated temperatures. *J. Membr. Sci.* **2014**, *450*, 407–417. [CrossRef]

12. Althuluth, M.; Overbeek, J.P.; van Wees, H.J.; Zubeir, L.F.; Haije, W.G.; Berrouk, A.; Peters, C.J.; Kroon, M.C. Natural gas purification using supported ionic liquid membrane. *J. Membr. Sci.* **2015**, *484*, 80–86. [CrossRef]

13. Anderson, J.L.; Dixon, J.K.; Brennecke, J.F. Solubility of CO_2, CH_4, C_2H_6, C_2H_4, O_2, and N_2 in 1-Hexyl-3-methylpyridinium Bis(trifluoromethylsulfonyl)imide: Comparison to Other Ionic Liquids. *Acc. Chem. Res.* **2007**, *40*, 1208–1216. [CrossRef] [PubMed]

14. Finotello, A.; Bara, J.E.; Camper, D.; Noble, R.D. Room-Temperature Ionic Liquids: Temperature Dependence of Gas Solubility, Selectivity. *Ind. Eng. Chem. Res.* **2008**, *47*, 3453–3459. [CrossRef]

15. Liu, X.; He, M.; Nan, L.; Xu, H.; Bai, L. Selective absorption of CO_2 from H_2, O_2 and N_2 by 1-hexyl-3-methylimidazolium tris(pentafluoroethyl) trifluorophosphate. *J. Chem. Thermodyn.* **2016**, *97*, 48–54. [CrossRef]

16. Afzal, W.; Yoo, B.; Prausnitz, J.M. Inert-Gas-Stripping Method for Measuring Solubilities of Sparingly Soluble Gases in Liquids. Solubilities of Some Gases in Protic Ionic Liquid 1-Butyl, 3-Hydrogen-imidazolium Acetate. *Ind. Eng. Chem. Res.* **2012**, *51*, 4433–4439. [CrossRef]

17. Sedov, I.A.; Magsumov, T.I.; Salikov, T.M.; Solomonov, B.N. Solvation of apolar compounds in protic ionic liquids: The non-synergistic effect of electrostatic interactions and hydrogen bonds. *Phys. Chem. Chem. Phys.* **2017**, *19*, 25352–25359. [CrossRef] [PubMed]

18. Alcantara, M.L.; Ferreira, P.I.S.; Pisoni, G.O.; Silva, A.K.; Cardozo-Filho, L.; Lião, L.M.; Pires, C.A.M.; Mattedi, S. High pressure vapor-liquid equilibria for binary protic ionic liquids + methane or carbon dioxide. *Sep. Purif. Technol.* **2018**, *196*, 32–40. [CrossRef]

19. Alcantara, M.L.; Santos, J.P.; Loreno, M.; Ferreira, P.I.S.; Paredes, M.L.L.; Cardozo-Filho, L.; Silva, A.K.; Lião, L.M.; Pires, C.A.M.; Mattedi, S. Low viscosity protic ionic liquid for CO_2/CH_4 separation: Thermophysical and high-pressure phase equilibria for diethylammonium butanoate. *Fluid Phase Equilib.* **2018**, *459*, 30–43. [CrossRef]

20. Lemus, J.; Eguizábal, A.; Pina, M.P. UV polymerization of room temperature ionic liquids for high temperature PEMs: Study of ionic moieties and crosslinking effects. *Inter. J. Hydrog. Energy* **2015**, *40*, 5416–5424. [CrossRef]

21. Lemus, J.; Eguizábal, A.; Pina, M.P. Endurance strategies for the preparation of high temperature polymer electrolyte membranes by UV polymerization of 1-H-3-vinylimidazolium bis(trifluoromethanesulfonyl)imide for fuel cell applications. *Inter. J. Hydrog. Energy* **2016**, *41*, 3981–3993. [CrossRef]

22. Eguizábal, A.; Lemus, J.; Roda, V.; Urbiztondo, M.; Barreras, F.; Pina, M.P. Nanostructured electrolyte membranes based on zeotypes, protic ionic liquids and porous PBI membranes: Preparation, characterization and MEA testing. *Inter. J. Hydrog. Energy* **2012**, *37*, 7221–7234. [CrossRef]

23. Kallem, P.E.A.; Mallada, R.; Pina, M.P. Constructing straight Poly-ionic liquid microchannels for fast anhydrous proton transport straight. *ACS Appl. Mater. Interfaces* **2016**, *8*, 35377–35389. [CrossRef] [PubMed]

24. Kallem, P.; Drobek, M.; Julbe, A.; Vriezekolk, E.J.; Mallada, R.; Pina, M.P. Hierarchical Porous Polybenzimidazole Microsieves: An Efficient Architecture for Anhydrous Proton Transport via Polyionic Liquids. *ACS Appl. Mater. Interfaces* **2017**, *9*, 14844–14857. [CrossRef] [PubMed]

25. Ferguson, L.; Scovazzo, P. Solubility, Diffusivity, and Permeability of Gases in Phosphonium-Based Room Temperature Ionic Liquids: Data and Correlations. *Ind. Eng. Chem. Res.* **2007**, *46*, 1369–1374. [CrossRef]

26. Scovazzo, P.; Havard, D.; McShea, M.; Mixon, S.; Morgan, D. Long-term, continuous mixed-gas dry fed CO_2/CH_4 and CO_2/N_2 separation performance and selectivities for room temperature ionic liquid membranes. *J. Membr. Sci.* **2009**, *327*, 41–48. [CrossRef]

27. Neves, L.A.; Crespo, J.G.; Coelhoso, I.M. Gas permeation studies in supported ionic liquid membranes. *J. Membr. Sci.* **2010**, *357*, 160–170. [CrossRef]

28. Cserjési, P.; Nemestóthy, N.; Bélafi-Bakó, K. Gas separation properties of supported liquid membranes prepared with unconventional ionic liquids. *J. Memb. Sci.* **2010**, *349*, 6–11. [CrossRef]

29. Tomé, L.C.; Mecerreyes, D.; Freire, C.S.R.; Rebelo, L.P.N.; Marrucho, I.M. Pyrrolidinium-based polymeric ionic liquid materials: New perspectives for CO_2 separation membranes. *J. Membr. Sci.* **2013**, *428*, 260–266. [CrossRef]

30. Borjigin, H.; Stevens, K.A.; Liu, R.; Moon, J.D.; Shaver, A.T.; Swinnea, S.; Freeman, B.D.; Riffle, J.S.; McGrath, J.E. Synthesis and characterization of polybenzimidazoles derived from tetraaminodiphenylsulfone for high temperature gas separation membranes. *Polymer* **2015**, *71*, 135–142. [CrossRef]

31. Han, S.H.; Lee, J.E.; Lee, K.-J.; Park, H.B.; Lee, Y.M. Highly gas permeable and microporous polybenzimidazole membrane by thermal rearrangement. *J. Membr. Sci.* **2010**, *357*, 143–151. [CrossRef]

32. Pesiri, D.R.; Jorgensen, B.; Dye, R.C. Thermal optimization of polybenzimidazole meniscus membranes for the separation of hydrogen, methane, and carbon dioxide. *J. Membr. Sci.* **2003**, *218*, 11–18. [CrossRef]

33. Sun, X.; Zhang, M.; Guo, R.; Luo, J.; Li, J. CO_2-facilitated transport performance of poly(ionic liquids) in supported liquid membranes. *J. Mater. Sci.* **2014**, *50*, 104–111. [CrossRef]

34. Hu, X.D.; Tang, J.B.; Blasig, A.; Shen, Y.Q.; Radosz, M. CO_2 permeability, diffusivity and solubility in polyethylene glycol-grafted polyionic membranes and their CO_2 selectivity relative to methane and nitrogen. *J. Membr. Sci.* **2006**, *281*, 130–138. [CrossRef]

35. Cera-Manjares, A.; Salavera, D.; Coronas, A. Vapour pressure measurements of ammonia/ionic liquids mixtures as suitable alternative working fluids for absorption refrigeration technology. *Fluid Phase Equilib.* **2018**, in press. [CrossRef]

36. Charmette, C.; Sanchez, J.; Gramain, P.; Rudatsikira, A. Gas transport properties of poly(ethylene oxide-co-epichlorohydrin) membranes. *J. Membr. Sci.* **2004**, *230*, 161–169. [CrossRef]

37. Howlett, P.C.; Brack, N.; Hollenkamp, A.F.; Forsyth, M.; MacFarlane, D.R. Characterization of the Lithium Surface in N-Methyl-N-alkylpyrrolidinium Bis(trifluoromethanesulfonyl)amide Room-Temperature Ionic Liquid Electrolytes. *J. Electrochem. Soc.* **2006**, *153*, A595. [CrossRef]

38. Carlisle, T.K.; Bara, J.E.; Gabriel, C.J.; Noble, R.D.; Gin, D.L. Interpretation of CO_2 Solubility and Selectivity in Nitrile-Functionalized Room-Temperature Ionic Liquids Using a Group Contribution Approach. *Ind. Eng. Chem. Res.* **2008**, *47*, 7005–7012. [CrossRef]

39. Young, J.S.; Long, G.S.; Espinoza, B.F. Cross-Linked Polybenzimidazole Membrane for Gas Separation. Google Patents US6946015B2, 20 September 2005.

40. Moschovi, A.M.; Ntais, S.; Dracopoulos, V.; Nikolakis, V. Vibrational spectroscopic study of the protic ionic liquid 1-H-3-methylimidazolium bis(trifluoromethanesulfonyl)imide. *Vib. Spectrosc.* **2012**, *63*, 350–359. [CrossRef]

41. Camper, D.; Bara, J.; Koval, C.; Noble, R. Bulk-Fluid Solubility and Membrane Feasibility of Rmim-Based Room-Temperature Ionic Liquids. *Ind. Eng. Chem. Res.* **2006**, *45*, 6279–6283. [CrossRef]

42. Condemarin, R.; Scovazzo, P. Gas permeabilities, solubilities, diffusivities, and diffusivity correlations for ammonium-based room temperature ionic liquids with comparison to imidazolium and phosphonium RTIL data. *Chem. Eng. J.* **2009**, *147*, 51–57. [CrossRef]

43. Scovazzo, P.; Kieft, J.; Finan, D.; Koval, C.; Dubois, D.; Noble, R. Gas separations using non-hexafluorophosphate [PF$_6$]−anion supported ionic liquid membranes. *J. Membr. Sci.* **2004**, *238*, 57–63. [CrossRef]

MDPI

St. Alban-Anlage 66

4052 Basel

Switzerland

Tel. +41 61 683 77 34

Fax +41 61 302 89 18

www.mdpi.com

Membranes Editorial Office

E-mail: membranes@mdpi.com

www.mdpi.com/journal/membranes

www.ingramcontent.com/pod-product-compliance
Lightning Source LLC
Chambersburg PA
CBHW051908210326
41597CB00033B/6073